ELECTROMAGNETICS FOR ENGINEERS

VOLUME 1

Electrostatics and Magnetostatics

For a complete listing of titles in the
Artech House Electromagnetic Analysis Library,
turn to the back of this book.

ELECTROMAGNETICS FOR ENGINEERS

VOLUME 1

Electrostatics and Magnetostatics

Dean James Friesen

ARTECH HOUSE
BOSTON | LONDON
artechhouse.com

Library of Congress Cataloging-in-Publication Data
A catalog record for this book is available from the U.S. Library of Congress.

British Library Cataloguing in Publication Data
A catalogue record for this book is available from the British Library.

ISBN 13: 978-1-68569-005-2

Cover design by Joi Garron

Artech House
685 Canton Street
Norwood, MA 02062

10 9 8 7 6 5 4 3 2 1

CONTENTS

6
CURRENT, CURRENT DENSITY, AND CONDUCTORS 73

7
CAPACITANCE AND DIELECTRIC MATERIALS 93

8

LAPLACE'S EQUATION 113

Part 2

MAGNETOSTATICS: BASIC PRINCIPLES 129

9

AMPERE'S LAW AND THE MAGNETIC FIELD 131

10

FORCES AND TORQUES IN MAGNETIC FIELDS 153

11

INDUCTANCE AND MAGNETIC CIRCUITS 171

12

DISPLACEMENT CURRENT AND INDUCED EMF 191

PART 1

ELECTROSTATICS: BASIC PRINCIPLES

1

VECTOR ANALYSIS

Electromagnetics principles can be described adequately with precise and well-crafted words. However, in the practical engineering world, there often arises the need to quantify the magnitude of an important and unsolved electromagnetic engineering problem—and effectively convey it to senior engineering and management. For such often unwelcome occasions, vector calculus comes in handy, seeing how it helped form (by way of many predecessors such as James Clerk Maxwell), the foundation of electromagnetics, more specifically, electrostatics and magnetostatics. Vector calculus is indispensable to the self-study and successful acquisition of a thorough understanding of the subject(s).

For convenience, a short refresher on vector calculus is offered in this chapter. The author nonetheless advises readers to acquire for themselves a very good and easy-to-comprehend calculus book and emphasizes using it to strengthen the reader's skill. It will make understanding the underlying principles of electromagnetics far easier. This simple overview also aids in keeping the size of this first volume to a manageable level for the publisher. All said, a highly recommended university/college-level, self-study calculus textbook is cited in [1].

1.1 INTRODUCTION

Discussion in this first chapter begins with a brief tutorial on vectors. Relevant coordinate systems are presented as are discussions on vector notation and structure, and the parallelogram and head-to-tail rules (see Appendix C). The unit vector is presented as: (1) differential volume, (2) differential surface (area), and (3) differential line elements. Discussion on vector fields and coordinate system transformations are also given.

1.2 VECTOR NOTATION

It is vital for the engineering analyst to be able to distinguish between vector quantities and scalar quantities. Vectors are quantities that have both a magnitude and a direction. Scalars are quantities that have only a magnitude; that is, no indicated direction [1]. Vectors throughout this text are indicated by boldface type. Scalars are denoted in plain type.

Distance formula is the distance between two points in three-dimensional space (Cartesians coordinates). It is a scalar value that can be found, given two coordinate points in space, by the following formula:

$$D = \left[\left(x_1 - x_0\right)^2 + \left(y_1 - y_0\right)^2 + \left(z_1 - z_0\right)^2 \right]^{1/2} \tag{1.1}$$

where

P_0 (x_0, y_0, z_0) and P_1 (x_1, y_1, z_1) are points in space [2].

A unit vector is always a vector having an absolute value (i.e., a magnitude or length) of 1 and a direction [3]. It will always be indicated in this text by boldface, lower case letters, such as, **a** or **u**. The unit vector in the direction of a vector **A** is determined by dividing **A** by its absolute value (or vector length) as shown below:

$$\mathbf{a}_A = \frac{\mathbf{A}}{|\mathbf{A}|} \text{ or } \frac{\mathbf{A}}{A} \tag{1.2}$$

where

$|\mathbf{A}| = A = (A_x{}^2 + A_y{}^2 + A_z{}^2)^{1/2}$

A_x is the magnitude of the **A** vector in the x-direction

A_y is the magnitude of the **A** vector in the y-direction

A_z is the magnitude of the **A** vector in the z-direction

It should be stated that many calculus texts use the boldfaced variable **u** to denote the unit vector as well [2].

Using unit vectors, $\mathbf{a}_x$, $\mathbf{a}_y$, and $\mathbf{a}_z$ along the Cartesian *x*, *y*, and *z* system coordinate axes, an arbitrary vector can be written in component form as:

$$\mathbf{A} = A_x\mathbf{a}_x + A_y\mathbf{a}_y + A_z\mathbf{a}_z = A_x\mathbf{x} + A_y\mathbf{y} + A_z\mathbf{z} \tag{1.3}$$

However, for writing convenience and reduction of confusion, the above-expressed Cartesian-based unit vectors will be written in this text from here forward as **x**, **y**, and **z**, as opposed to $\mathbf{a}_x$, $\mathbf{a}_y$, and $\mathbf{a}_z$. The same parallel cases will also be for unit vectors for cylindrical and spherical coordinates [3].

Hence, unit vectors for the cylindrical coordinate system will be written as the more conveniently written

$$\mathbf{r}, \boldsymbol{\phi}, \text{ and } \mathbf{z}$$

instead of the more awkward mathematician's form of $\mathbf{a}_r$, $\mathbf{a}_\phi$, and $\mathbf{a}_z$, respectively. Likewise, the case for the spherical coordinate system; that is:

$$\mathbf{r}, \boldsymbol{\phi}, \text{ and } \boldsymbol{\theta}$$

for $\mathbf{a}_r$, $\mathbf{a}_\phi$, and $\mathbf{a}_\theta$ [1]. In determining the resultant vector $\mathbf{A} + \mathbf{B}$, that is in the case where two vectors follow each other in a head-to-tail connection with each other, the parallelogram rule is frequently employed. Because it is important to understand well the parallelogram rule for the solution of many electrostatics problems, a detailed discussion is offered in Appendix C of this book.

1.3 VECTOR ALGEBRA

1. Vectors may be added and subtracted.

$$\begin{aligned} \mathbf{A} \pm \mathbf{B} &= \left(A_x\mathbf{x} + A_y\mathbf{y} + A_z\mathbf{z}\right) \pm \left(B_x\mathbf{x} + B_y\mathbf{y} + B_z\mathbf{z}\right) \\ &= \left(A_x \pm B_x\right)\mathbf{x} + \left(A_y \pm B_y\right)\mathbf{y} + \left(A_z \pm B_z\right)\mathbf{z} \end{aligned} \tag{1.4}$$

2. The associative,

$$\mathbf{A} + (\mathbf{B} + \mathbf{C}) = (\mathbf{A} + \mathbf{B}) + \mathbf{C} \tag{1.5}$$

distributive,

$$k(\mathbf{A} + \mathbf{B}) = k\mathbf{A} + k\mathbf{B}, \quad (k_1 + k_2)\mathbf{A} = k_1\mathbf{A} + k_2\mathbf{A} \tag{1.6}$$

where
 k is an arbitrary constant of any value
 k_1 is an arbitrary constant of a particular value
 k_2 is a second arbitrary constant of a value not equal to k_1
and commutative laws apply.

$$\mathbf{A} + \mathbf{B} = \mathbf{B} + \mathbf{A} \tag{1.7}$$

3. The dot product of two vectors is, by definition:

$$\mathbf{A} \cdot \mathbf{B} = |\mathbf{A}|\,|\mathbf{B}|\cos\Theta = AB\cos\Theta \quad \text{(Note: This reads "A dot B")} \tag{1.8}$$

where Θ is the smaller angle between vectors **A** and **B**.

From the component form it can be shown that:

$$\mathbf{A} \cdot \mathbf{B} = A_x B_x + A_y B_y + A_z B_z \tag{1.9}$$

in particular,

$$\mathbf{A} \cdot \mathbf{A} = |\mathbf{A}|^2 = A_x^2 + A_y^2 + A_z^2 \tag{1.10}$$

4. The cross product of two vectors is, by definition:

$$\mathbf{A} \times \mathbf{B} = (|\mathbf{A}||\mathbf{B}|\sin\theta)\mathbf{n} = (AB\sin\Theta)\mathbf{n} \quad \text{(Reads "A cross B")} \tag{1.11}$$

where Θ is the smaller angle between vectors **A** and **B** and **n** is a unit vector normal to the plane determined by **A** and **B** when they are drawn from a point common to both [1].

In the cross product, two normals are in the plane [1], and, thus, further specification is required [4]. The selected normal,

$$\mathbf{A} \times \mathbf{B} = -\mathbf{B} \times \mathbf{A} \tag{1.12}$$

is the one that is in the direction of a right-hand screw when **A** is turned toward **B** (see Figure 1.1).

The cross product is anticommutative; that is, it does not follow the commutative principle [1]. Expanding the cross product into component form yields:

$$\begin{aligned}\mathbf{A} \times \mathbf{B} &= \left(A_x\mathbf{x} + A_y\mathbf{y} + A_z\mathbf{z}\right) \times \left(B_x\mathbf{x} + B_y\mathbf{y} + B_z\mathbf{z}\right) \\ &= \left(A_yB_z - A_zB_y\right)\mathbf{x} + \left(A_zB_x - A_xB_z\right)\mathbf{y} + \left(A_xB_y - A_yB_x\right)\mathbf{z} \\ &= \begin{bmatrix} \mathbf{x} & \mathbf{y} & \mathbf{z} \\ A_X & A_y & A_z \\ B_x & B_Y & B_z \end{bmatrix}\end{aligned} \tag{1.13}$$

conveniently expressed in determinant form (< immediately left).

The following is a triple scalar product of three vectors (**A**, **B**, and **C**):

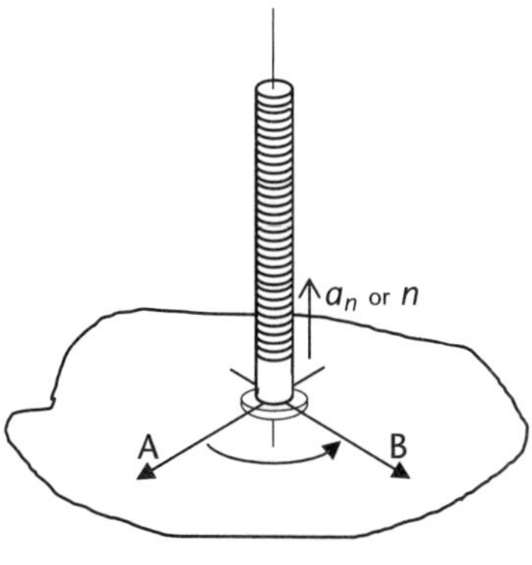

Figure 1.1 Cross-product illustration.

$$[\mathbf{A}\ \mathbf{B}\ \mathbf{C}] = \mathbf{A} \cdot (\mathbf{B} \times \mathbf{C}) = (\mathbf{A} \times \mathbf{B}) \cdot \mathbf{C} = \begin{bmatrix} A_1 & A_2 & A_3 \\ B_1 & B_2 & B_3 \\ C_1 & C_2 & C_3 \end{bmatrix} \tag{1.14}$$

Equation (1.14) yields the volume of a parallelepiped (see Figure 1.2) [1].

1.4 COORDINATE SYSTEMS

The Cartesian coordinate system, displayed in Figure 1.3(a), is one of three coordinate systems that are used to help solve electromagnetics-related engineering problems. The other two coordinate systems are the cylindrical coordinate system (see Figure 1.3(b)) and the spherical coordinate system (see Figure 1.3(c)) [1]. All three are displayed in the figure to give the reader the chance to visualize and understand similarities/differences between the coordinate systems.

As can be seen in Figure 1.3, the Cartesian and cylindrical coordinate systems have the z-axis in common. The cylindrical and spherical coordinate systems have the ϕ-coordinate in common. However, while these two systems also have an r-coordinate, each

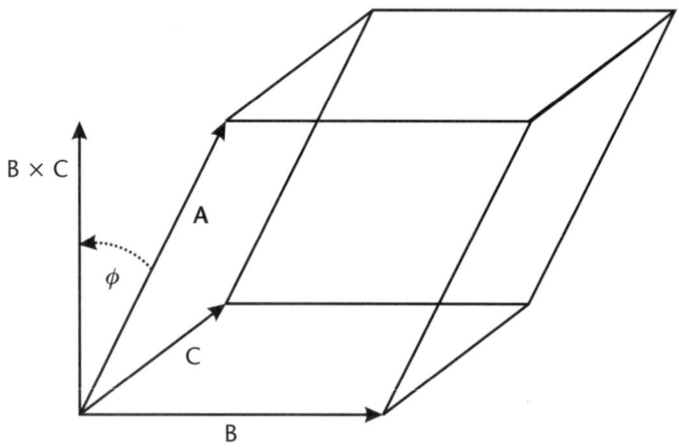

$$V_{parallelepiped} = |\mathrm{A} + (\mathrm{B} \times \mathrm{C})|$$
$$= |\mathrm{B} \times \mathrm{C}|\ |\mathrm{A}|\ |\cos \phi$$

Figure 1.2 Parallelepiped illustration with associated vector expressions.

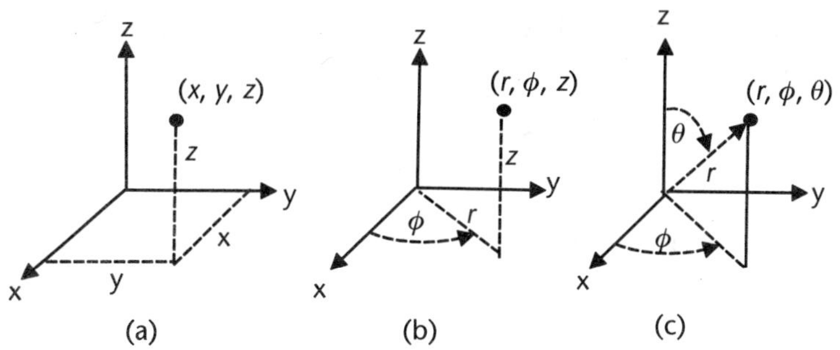

Figure 1.3 Three coordinate systems that are used to solve electromagnetic problems: (a) Cartesian, (b) cylindrical, and (c) spherical.

manifests itself in slightly different orientations. Hence, the r-coordinate for the spherical coordinate system appears to have slightly more directional freedom than the r-coordinate for the cylindrical coordinate.

Figure 1.4 provides a second helpful set of illustrations for all three coordinate systems:

- Figure 1.4(a)—the Cartesian coordinate system;
- Figure 1.4(b)—the cylindrical coordinate system;
- Figure 1.4(c)—the spherical coordinate system.

Each of the illustrations intends to show the geometric envelope of each system [1].

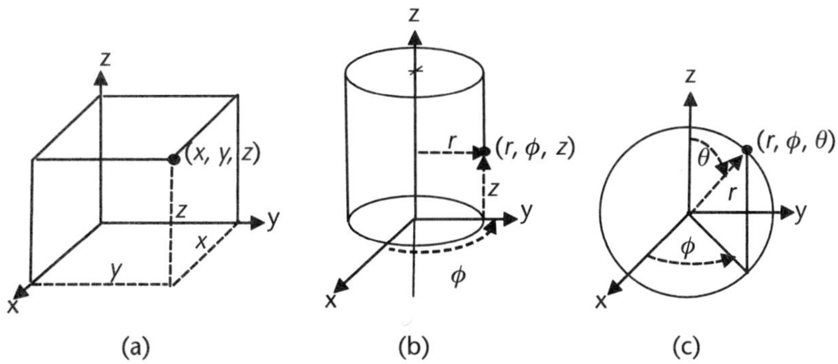

Figure 1.4 Illustrations of the use for each three-dimensional coordinate system: (a) Cartesian, (b) cylindrical, and (c) spherical.

Ranges of the coordinates (r, ϕ, z) in the cylindrical coordinate system are:

$$0 < r < \infty, \qquad 0 < \phi < 2\pi, \qquad -\infty < z < +\infty \tag{1.15}$$

Ranges of the coordinates (r, ϕ, θ) in the spherical coordinate system are:

$$0 < r < \infty, \qquad 0 < \phi < 2\pi, \qquad 0 < \theta < \pi \tag{1.16}$$

The Cartesian coordinate system, the one shown in Figures 1.3(a) and 1.4(a), is the one with which beginning engineers and engineering students are likely to have the greatest initial familiarity. However, in certain cases, solutions are sometimes found to be easier to obtain by using either the cylindrical or spherical coordinate system rather than the Cartesian [1].

Practical, real-world examples for the use of the cylindrical coordinate system include the geometries of wires, wire bundle harnesses, coaxial cables, and airplane fuel quantity sensors. In these cases, the cylindrical coordinate system lends itself more easily to solution-finding than does the Cartesian coordinate system. For the study of antennas in antenna reference texts, the reader will find that antenna performance and antenna patterns (E-field or power) are more easily evaluated in the spherical coordinate system.

In this text, the right-handed versions of the Cartesian, the cylindrical, and the spherical coordinate systems are used. (Why add to the confusion to use the left-handed version? It adds no value to the real-world of finding solutions to practical problems.) Hence:

$$\mathbf{x} \times \mathbf{y} = \mathbf{z} \tag{1.17}$$

$$\mathbf{r} \times \boldsymbol{\phi} = \mathbf{z} \tag{1.18}$$

$$\mathbf{r} \times \boldsymbol{\theta} = \boldsymbol{\phi} \tag{1.19}$$

The component forms of a vector, $\mathbf{A}$, in the three coordinate systems are:

$$\mathbf{A} = A_x\mathbf{x} + A_y\mathbf{y} + A_z\mathbf{z} \quad \text{(Cartesian)} \tag{1.20}$$

$$\mathbf{A} = A_r\mathbf{r} + A_\phi\boldsymbol{\phi} + A_z\mathbf{z} \quad \text{(cylindrical)} \tag{1.21}$$

$$\mathbf{A} = A_r\mathbf{r} + A_\phi\boldsymbol{\phi} + A_\theta\boldsymbol{\theta} \quad \text{(spherical)} \tag{1.22}$$

Most often, the components A_x, A_y, A_z, A_r, A_ϕ, and A_θ shown in (1.20) to (1.22), generally speaking, are not always constants. It is important to understand that each can also be *functions* in the coordinate system form as well [1].

1.5 DIFFERENTIAL VOLUME, SURFACE, AND LINE ELEMENTS

Differential Volume. When the coordinates of a point P are expanded to:

$$(x + dx, y + dy, z + dz) \quad \text{(Cartesian coordinate system)} \tag{1.23}$$

$$(r + dr, \phi + d\phi, z + dz) \quad \text{(cylindrical coordinate system)} \tag{1.24}$$

or

$$(r + dr, \theta + d\theta, \phi + d\phi) \quad \text{(spherical coordinate system)} \tag{1.25}$$

a differential volume, dV, is formed [1]. The expressions for each coordinate system are:

$$dV = dx\ dy\ dz \quad \text{(Cartesian coordinate system)} \tag{1.26}$$

$$dV = dr(r\ d\phi)dz = r\ dr\ d\phi\ dz \quad \text{(cylindrical coordinate system)} \tag{1.27}$$

$$dV = dr(r\ d\theta)(r\sin\theta\ d\phi) = r^2\ dr\sin\theta\ d\theta\ d\phi \quad \text{(spherical coordinate system)} \tag{1.28}$$

To the first order in infinitesimal quantities, the differential volume for the Cartesian coordinate system is a rectangular box (i.e., cuboid). See Figure 1.5(a) [1]. In the other two coordinate systems (see Figures 1.5(b) and 1.5(c)), the corresponding volumes are slightly distorted variations from the Cartesian cuboid.

The reader is encouraged to carefully study the latter two differential volumes until a good, accepted understanding of each is reached.

Differential surface (area). When a differential surface (area) is the focus for quantitative evaluation, the dS expressions for each of the three coordinate systems are used [6]. These are presented for convenience as follows:

$$dS = dx\ dy \quad dS = dy\ dz \quad dS = dx\ dz \quad \text{(Cartesian)} \tag{1.29}$$

$$\begin{aligned} dS = dr\ dz \quad dS &= dz(r\ d\phi) \quad dS = dr(r\ d\phi) \\ &= r\ d\phi\ dz \qquad\quad = (r\ dr)d\phi \end{aligned} \quad \text{(cylindrical)} \tag{1.30}$$

$$\begin{aligned} dS &= dr(r\ d\theta) \quad dS = dr(r\sin\theta\ d\phi) \quad dS = (r\ d\theta)(r\sin\theta\ d\phi) \\ &= r\ dr\ d\phi \qquad\qquad\qquad\qquad\qquad\quad = r^2\sin\theta\ d\theta\ d\phi \end{aligned} \quad \text{(spherical)} \tag{1.31}$$

The reader is urged to also use Figure 1.5 as an aid to gaining visual understanding of differential surfaces on each of the differential volumes as well.

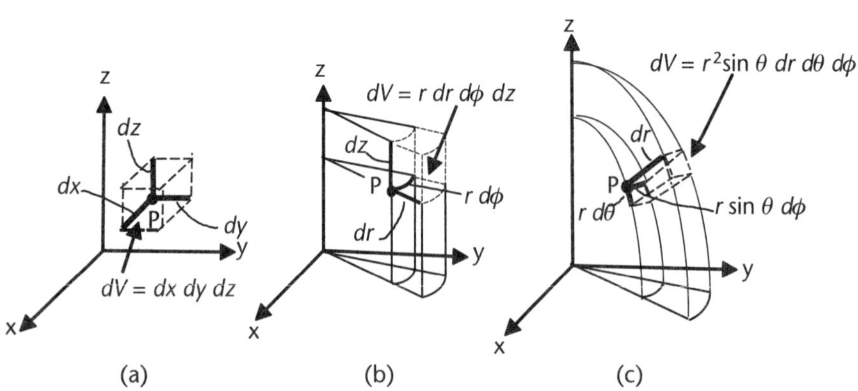

Figure 1.5 Differential volume, *dV*, in each of three coordinate systems: (a) Cartesian, (b) cylindrical, and (c) spherical.

Differential line element. The differential line, *dl*, is the diagonal through an arbitrary point P [5].

Expressions for the differential line elements that correspond to each of the coordinate systems are given as follows [5]:

$$dl = dx^2 + dy^2 + dz^2 \qquad \text{(Cartesian)} \tag{1.32}$$

$$dl = dr^2 + r^2 d\phi^2 + dz^2 \qquad \text{(cylindrical)} \tag{1.33}$$

$$dl = dr^2 + r^2 d\theta^2 + r^2 \sin^2\theta d\phi^2 \qquad \text{(spherical)} \tag{1.34}$$

It is vital for the reader to understand that the above formulas for differential volume, differential surface, and differential line element will be fundamental to quantitative problem-solving in Chapters 2 to 8.

1.6 VECTOR FIELDS

A *field* is a spatial distribution of a quantity that may or may not be a function of time. Generally speaking, coefficients of electromagnetic vector expressions contain variables. As such, the expression changes magnitude and direction from point to point throughout the region of interest [1]. For example, consider the vector:

$$\mathbf{E} = -x\mathbf{x} + y\mathbf{y} \tag{1.35}$$

Values of x and y may be substituted into the expression to give **E** at the various locations.

After a number of points are examined, the pattern becomes evident, and Figure 1.6 shows this field.

A vector field may also vary with time, and, thus, the two-dimensional field examined in Figure 1.6 may also change with time. It could be assigned a time-variation such as:

$$\mathbf{E} = \left(-x\mathbf{a}_x + y\mathbf{a}_y\right)\sin(\omega t) \text{ or } E = \left(-x\mathbf{a}_x + y\mathbf{a}_x\right)e^{j\omega t} \tag{1.36}$$

The electric and magnetic fields of Chapter 2 and beyond are all time-variable. As might be expected, they will need to be

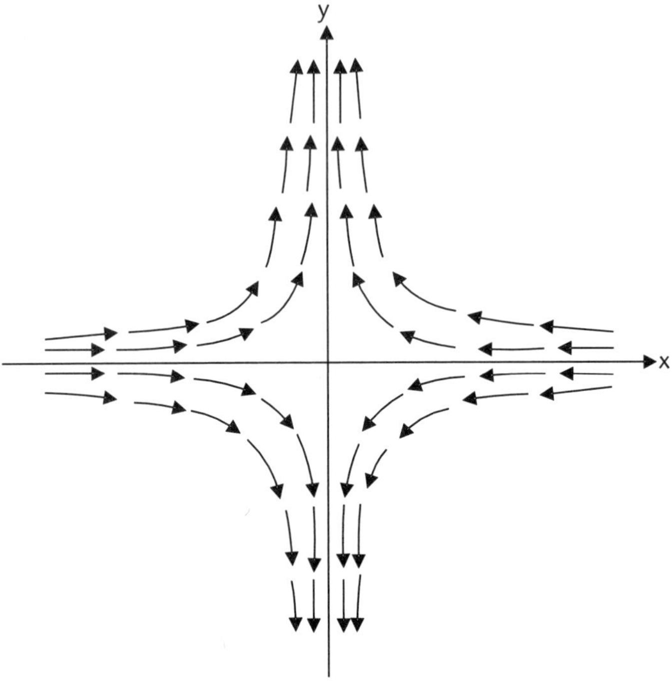

Figure 1.6 Example of a vector field, illustrated.

differentiated with respect to time and integrated with respect to time as the solution situation dictates. However, in time the reader will find that both operations follow naturally and rarely cause great difficulty [1].

If the reader believes he or she would benefit from vector analysis practice, he or she is encouraged to go to Section 1.8 where sample vector analysis problems are performed.

1.7 TRANSFORMATIONS BETWEEN COORDINATE SYSTEMS

Sometimes, it is helpful to transform points in space to one or the other of the two coordinate systems in order to obtain a solution more quickly. Here are transformations [1]:

Cartesian to cylindrical coordinate system. To deal with problems having cylindrical electromagnetic structures, the cylindrical coordinate system is used. If the reader finds it helpful to transform Cartesian coordinate points in space to equivalent cylindrical coordinate points, then the following transformational equations are used:

$$r = \left(x^2 + y^2\right)^{1/2} \qquad \phi = \arctan\left(\frac{y}{x}\right) \qquad z = z \tag{1.37}$$

Reverse cylindrical to Cartesian transformational equations are:

$$x = r\cos\phi \qquad y = r\sin\phi \qquad z = z \tag{1.38}$$

Cartesian to spherical coordinate system. To deal with problems having spherical electromagnetic structures, the spherical coordinate system is used. If the reader finds it helpful to transform a Cartesian coordinate point to an equivalent spherical coordinate point, then the following transformational equations are employed:

$$r = \left(x^2 + y^2 + z^2\right)^{1/2} \qquad \theta = \arctan\left[\frac{z}{\left(x^2 + y^2 + z^2\right)^{1/2}}\right] \qquad \phi = \arctan\left(\frac{y}{x}\right) \tag{1.39}$$

Reverse spherical to Cartesian transformational equations are:

$$x = r\sin\theta\cos\phi \qquad y = r\sin\theta\sin\phi \qquad z = r\cos\theta \tag{1.40}$$

1.8 PROBLEMS AND SOLUTIONS: VECTOR ANALYSIS

Problem 1.1

See Figure 1.7. Find the expression for vector **C** that is directed from point Q_1 (x_1, y_1, z_1) to Q_2 (x_2, y_2, z_2).

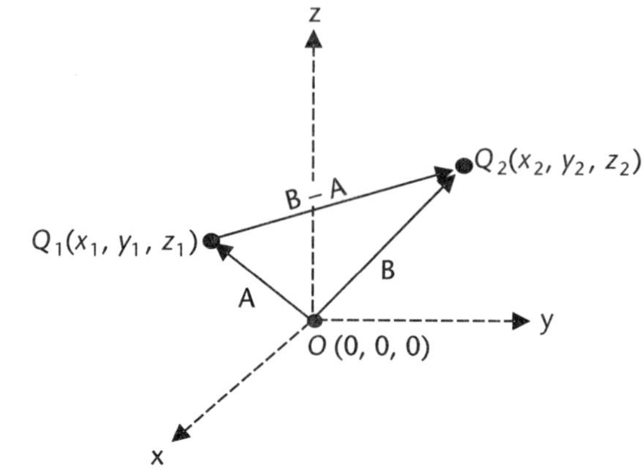

Figure 1.7 Vector diagram illustration for Problem 1.1.

Problem 1.2

See Figure 1.8. Find:

(a) The expression for vector **A** when Q_1 is located at (2, −4, 1) to Q_2 is located at (0, −2, 0) in Cartesian coordinates;

(b) The unit vector along vector **A**.

Problem 1.3

See Figure 1.9. Find the distance between the points $P_0[0, -5, 0]$ and $P_1(0, 5, 10)$ in Cartesian coordinates.

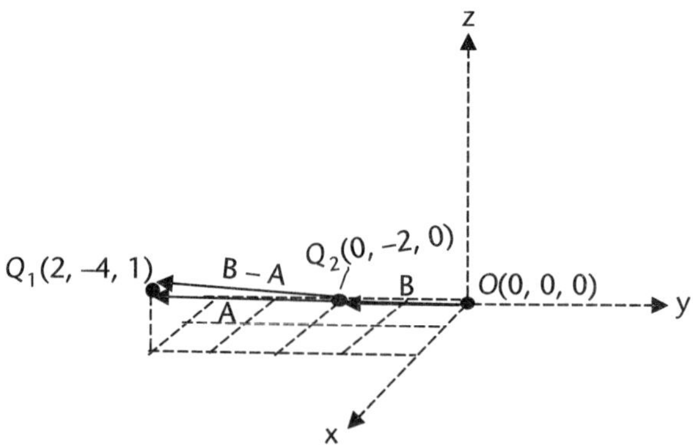

Figure 1.8 Vector diagram illustration for Problem 1.2.

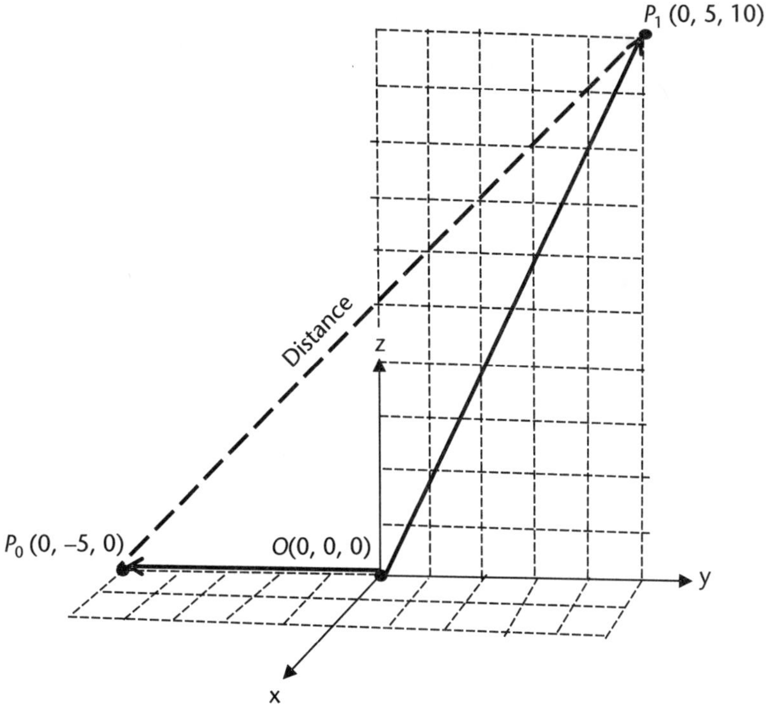

Figure 1.9 Vector diagram illustration for Problem 1.3.

Problem 1.4

See Figure 1.10. Given $\mathbf{A} = 2x + 4y - 3z$ and $\mathbf{B} = x - y$, find:

(a) $\mathbf{A} \cdot \mathbf{B}$,
(b) $\mathbf{A} \times \mathbf{B}$.

Problem 1.5

Show that:

$$\mathbf{A} = 4\mathbf{x} - 2\mathbf{y} - \mathbf{z} \quad \& \quad \mathbf{B} = \mathbf{x} + 4\mathbf{y} - 4\mathbf{z}$$

are perpendicular.

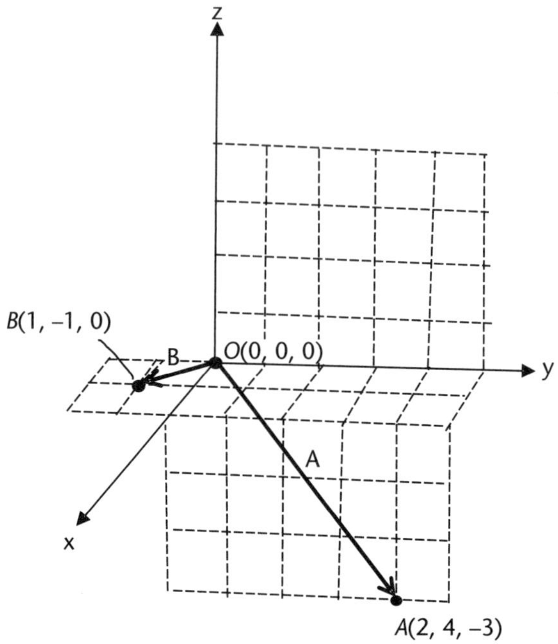

Figure 1.10 Vector diagram illustration for Problem 1.4.

Problem 1.6

Given that: $\mathbf{A} = 2\mathbf{x} + 4\mathbf{y}$ and $\mathbf{B} = 6\mathbf{y} - 4\mathbf{z}$, find the smaller angle between the vectors using the cross product and the dot product.

Problem 1.7

Given $\mathbf{F} = (y - 1)\mathbf{x} + (2x)\mathbf{y}$, find:

- The vector at (2, 2, 1),
- Its projection on B where $B = 5\mathbf{x} - \mathbf{y} + 2\mathbf{z}$.

Problem 1.8

Given:

$\mathbf{A} = \mathbf{x} + \mathbf{y}$,

$\mathbf{B} = \mathbf{x} + 2\mathbf{z}$,

$\mathbf{C} = 2\mathbf{y} + \mathbf{z}$,

find $(\mathbf{A} \times \mathbf{B}) \times \mathbf{C}$ and $\mathbf{A} \times (\mathbf{B} \times \mathbf{C})$.

Compare the answers.

Problem 1.9

Given:

$$\mathbf{A} = \mathbf{x} + \mathbf{y},$$
$$\mathbf{B} = \mathbf{x} + 2\mathbf{z},$$
$$\mathbf{C} = 2\mathbf{y} + \mathbf{z},$$

find $\mathbf{A} \cdot \mathbf{B} \times \mathbf{C}$, and find $\mathbf{A} \times \mathbf{B} \cdot \mathbf{C}$.

Compare the answers.

Problem 1.10

Express the unit vector that points from P_1 $(z_1 = h)$ on the z-axis toward $P_o(r, \phi, z_o)$, where $z_o = 0$, in cylindrical coordinates. See Figure 1.11.

Problem 1.11

Express the unit vector that is *directed toward the origin* from an arbitrary point on the plane, $z = -5$. Examine Figure 1.12.

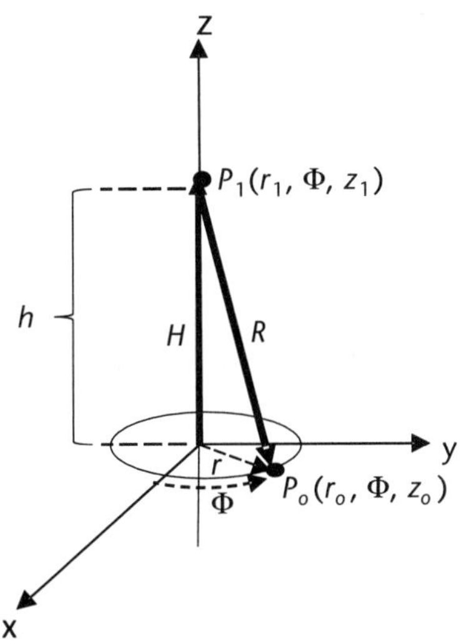

Figure 1.11 Vector diagram illustration for Problem 1.10.

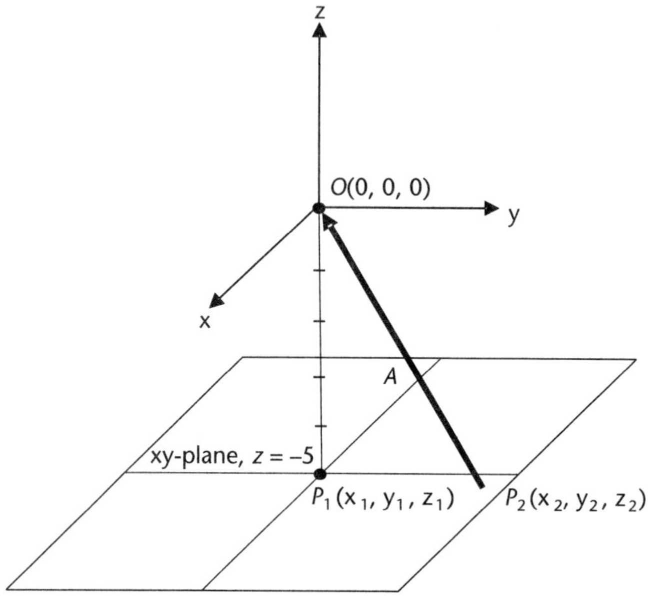

Figure 1.12 Vector diagram illustration for Problem 1.11.

Problem 1.12

Using the spherical coordinate system,

(a) Find the surface area of a strip $\alpha \leq \Theta \leq \beta$ on the spherical shell of radius α (see Figure 1.13).

(b) What is the result (area) when $\alpha = 0$ and $\beta = \pi$?

Problem 1.13

Develop the equation for the volume of a sphere of radius α from the differential volume equation.

Problem 1.14

Use the cylindrical coordinate system to find the area of the curved surface of a right circular cylinder where $r = 2\text{m}$, $h = 5$, and $30° < \phi < 120°$ (see Figure 1.14).

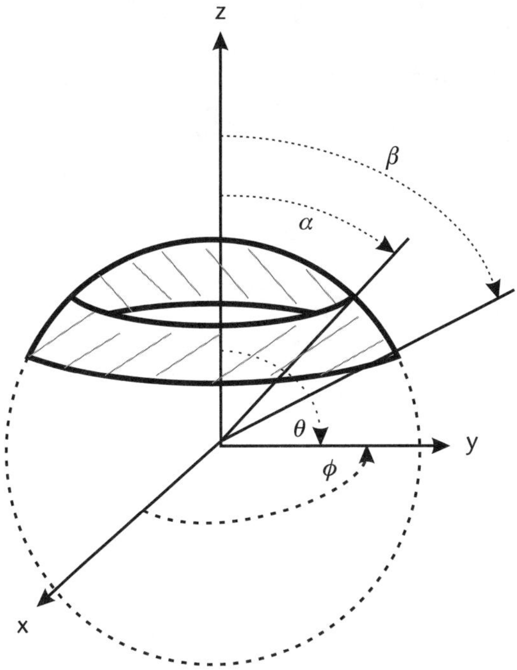

Figure 1.13 Vector diagram illustration for Problem 1.12.

Problem 1.15

Examine the cylinder of Figure 1.15. Quantify the surface areas of surfaces #1, #2, and #3 in the figure if the height of the cylinder is 10m, the inner cylinder radius, r_1, is 2m, and the outer cylinder radius, r_2, is 5m.

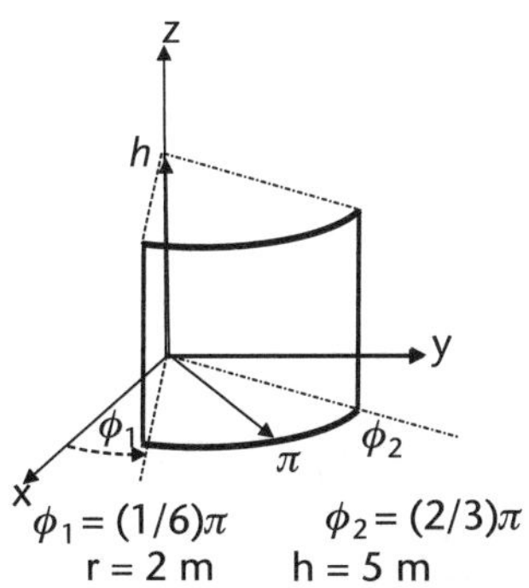

Figure 1.14 Vector diagram illustration for Problem 1.14.

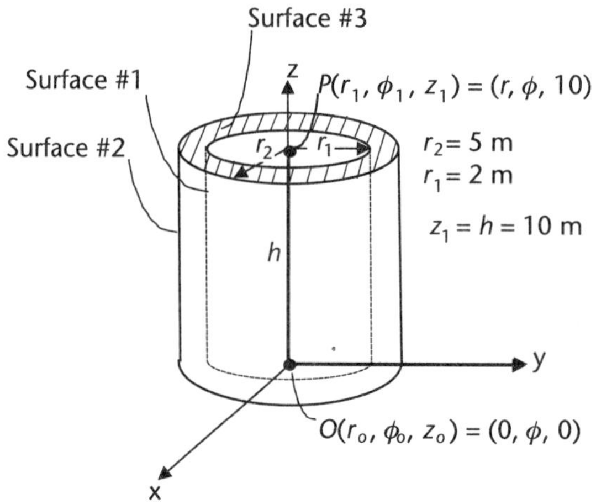

Figure 1.15 Vector diagram illustration for Problem 1.15.

References

[1] Boyce, W. E., *Calculus*, New York: John Wiley & Sons, 1988.

[2] Munein, M., and D. Foulis, *Calculus with Analytic Geometry*, New York: Worth Publishers, Inc., 1978.

[3] Anton, H., *Calculus with Analytic Geometry, Second Edition*, New York: John Wiley & Sons (Anton Textbooks, Inc.), 1984.

[4] Macnamara, T., *Handbook of Antennas for EMC*, Norwood, MA: Artech House, 1995.

[5] Stewart, J., *Calculus, Eighth Edition*, Boston, MA: Cengage Learning, 2016.

[6] Ayres, Jr., F., *Theory and Problems of Differential and Integral Calculus*, Second Edition, New York: McGraw-Hill, 1978.

2

COULOMB FORCES AND ELECTRIC FIELD INTENSITY

Stationery electrostatic charge generation occurs every day, and it is everywhere. It can be demonstrated to exist by way of:

- Walking inside a house on a room carpet in rubber-sole shoes on a dry-air wintry day and touching a metal doorknob;
- The pulling of Saran Wrap off the cylinder to cover a plastic salad dish (hence, the term "static cling");
- The removal of clothes from a warm dryer and noticing how, when you pull them apart, you see/hear arcs/sparks;
- The combing of dry hair in a dry winter climate using a rubber/plastic comb.

Other, less commonly known examples found in the real-world include static charge generation on the body and wings of a commercial jet (test) airplane occurring by flying above a rainstorm cloud at a (subzero temperature) altitude during a *p*-static flight test. During the test flight, the plane accumulates electrostatic (known as precipitation static) charge as evidenced by the need to bleed off the high charge remaining on the airplane after landing, parking, and using a large low-AWG conductive cable to bleed the accumulated charge off.

It may be of side interest for the reader to know that, of the four forces in the physical world, (i.e., nuclear, electrostatic, magnetostatic, and gravitational), electrostatic force is quite powerful and is the second most powerful force in the universe.

The fundamentals of electrostatics will be explored in this chapter along with introducing associated laws and equations. It should be noted that the focus of the next five chapters is on static charge, that is, charge that is not moving or changing in magnitude.

2.1 COULOMB'S LAW

Two electric charges, whether like or unlike, exert a force one on the other. If the charges are like-charges, the force is a *force of repulsion*. If the charges are unlike, the force is a *force of attraction*. Whichever the type, the force is directly proportional to the magnitude of each charge, Q_1 and Q_2, and the force is inversely proportional to the square of the separation distance between the charges. This is Coulomb's law. It was developed from direct observation of small, charged body interaction as made possible by use of a delicate torsion balance.

Throughout this text, calculations will be made in SI units. As such, units for repulsion or attraction force are in units of newtons (N). Units for the separation distance, d, between the charges are given in meters (m). The derived unit of charge is the coulomb (C). Force determination is rationalized by the product, 4π, and will be introduced via (2.3).

ε is the medium permittivity given with the units of C^2/Nm^2 or, equivalently (and, more commonly), F/m. Free-space permittivity (vacuum) is:

$$\varepsilon_0 = 8.854 \times 10^{-12} \quad \text{(F/m)} \tag{2.1}$$

For dielectric media other than free space, permittivity is a product of a material-specific dielectric constant, ε_r, and free-space permittivity, ε_0:

$$\varepsilon = \varepsilon_0 \varepsilon_r \quad \text{(F/m)} \tag{2.2}$$

where ε_r = *relative permittivity* or *dielectric constant of the media.*

In all solved problems throughout the text, it is important to know that free-space conditions are assumed unless otherwise so indicated. In vector form, Coulomb's law is stated as:

$$\mathbf{F} = \left(\frac{Q_1 Q_2}{4\pi\varepsilon d^2}\right)\mathbf{u},\ \text{(N)} \tag{2.3}$$

where

Q_1 = charge level of Pt 1 (C);

Q_2 = charge level of Pt 2 (C);

ε = dielectric material (or free space) permittivity between charges (F/m);

d = separation distance between charge points in space (m);

$\mathbf{u} = \mathbf{r}_{21}$ unit vector.

Subscripts on the unit vector (example, $\mathbf{r}_{21}$), help identify the force and indicate force direction:

$$\mathbf{F}_{21} = \left(\frac{Q_1 Q_2}{4\pi\varepsilon_0 d^2}\right)\mathbf{u}_{21}\ \text{(N)} \tag{2.4}$$

Equation (2.4) describes the force, $\mathbf{F}_{21}$, *on* Q_1 *from* Q_2. Convention is that the notation $\mathbf{u}_{21}$ directs the vector from Q_2 to Q_1 (see Figure 2.1).

In the region around an isolated point charge, there is a spherically symmetrical force field that is made evident when a charge, Q, is fixed at the origin and charges of equal magnitude, Q_1 through Q_n, and like sign are positioned volumetrically around the origin (see Figure 2.2).

This can also be expressed in spherical coordinates given as follows:

$$\mathbf{F} = \left(\frac{Q_1 Q_2}{4\pi\varepsilon_0 r^2}\right)\mathbf{r} = \left(\frac{Q_1 Q_2}{4\pi\varepsilon_0 R^2}\right)\mathbf{r}\ \text{(N)} \tag{2.5}$$

From a review of Figure 2.3, it can be seen that the symmetrical field about Q_o is disturbed by a second charge Q_1.

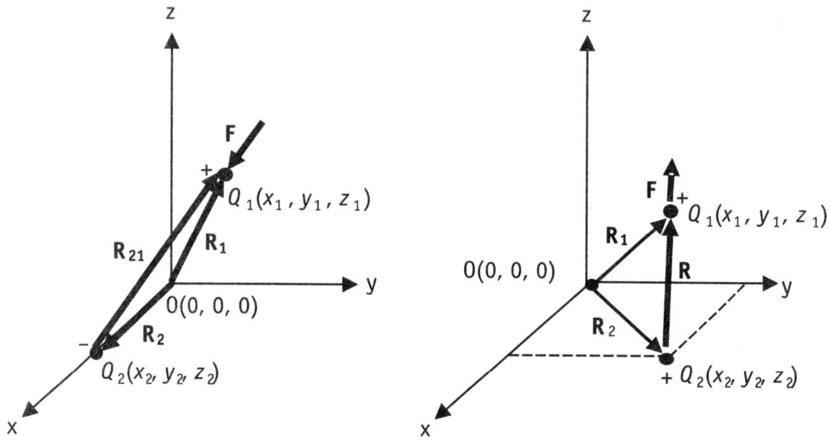

Figure 2.1 Two cases of a vector force, **F**, on a charge from another charge.

The resulting force, $\mathbf{F}_{\text{result}}$, is the vector sum of the two forces, $\mathbf{F}_0$ and $\mathbf{F}_1$, from charges Q_0 and Q_1,

$$\mathbf{F}_{\text{result}} = \mathbf{F}_0 + \mathbf{F}_1 \quad \text{(N)} \tag{2.6}$$

Likewise, just as there is a spherical force field around Q_0, there is also a spherical force field that is around Q_1, too. The resulting field between the two fields is a point-by-point vector sum, that is, the *superposition principle* for coulomb forces, and extends to any number of charges [1].

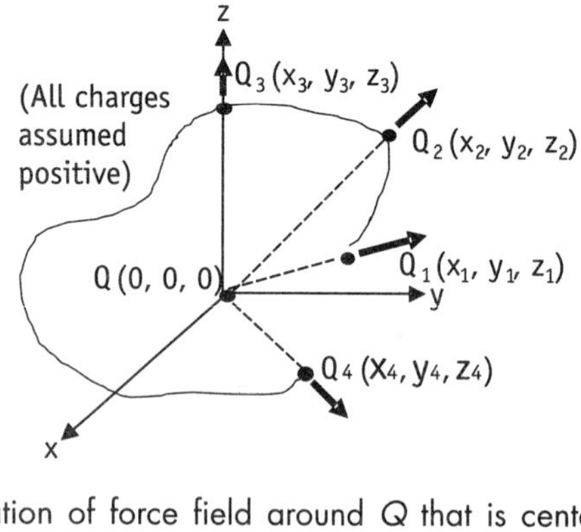

Figure 2.2 Illustration of force field around Q that is centered at the Cartesian coordinate system origin.

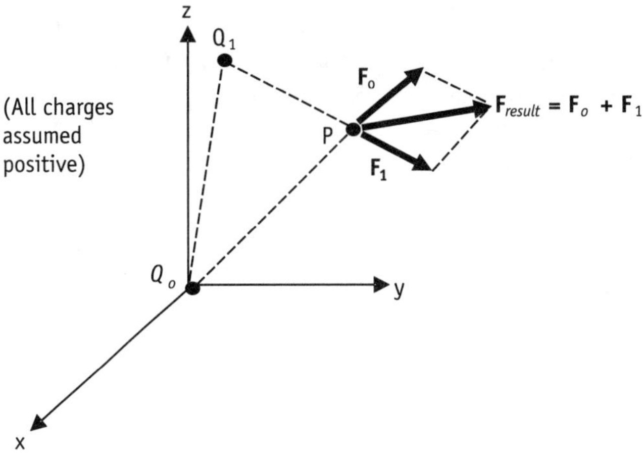

Figure 2.3 Illustration of electric forces together acting on a point in space, vector force sum result.

2.2 ELECTRIC FIELD INTENSITY

Suppose that in Figure 2.3, charge Q_1 is so small that it does not significantly disturb the fixed charge point Q_o (see Figure 2.4). In this case, the electric field (*E*-field) intensity, **E**, due to Q_o is defined to be the force per unit charge on Q_1, that is:

$$\mathbf{E} = \left(\frac{1}{Q_1}\right)\mathbf{F}_1 = \left(\frac{Q_0}{4\pi\varepsilon_0 R^2}\right)\mathbf{r}_1 \quad \text{(V/m)} \tag{2.7}$$

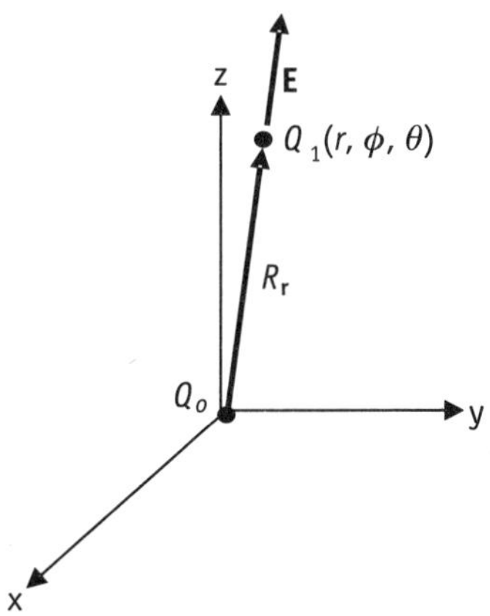

Figure 2.4 Illustration of electric field intensity and its vector (spherical coordinates).

In (2.7), **E** is given in spherical coordinates. The location of Q_o is placed at the origin. It may be transformed to other coordinate systems by the method described in Section 1.6.

In an arbitrary Cartesian coordinate system (see Figure 2.5), the equation for **E** may be expressed as follows:

$$\mathbf{E} = \left(\frac{Q}{4\pi\varepsilon_0 R^2}\right)\mathbf{R} \quad \text{(N/C or V/m)} \tag{2.8}$$

where the separation vector for R is **R**:

$$\mathbf{R} = (x_2 - x_1)\mathbf{x} + (y_2 - y_1)\mathbf{y} + (z_2 - z_1)\mathbf{z}.$$

The units for **E** are newtons per coulomb (N/C) as used by physicists. The more commonly used unit in the EM engineering field is volts per meter (V/m) [2].

2.3 CHARGE DISTRIBUTIONS

Up to this point, charge has been discussed in terms of a point or a set of points in space. There are other geometric distributions of charge that occur in the physical world.

Volume charge. When electric charge is distributed throughout a specified volume, each element of charge contributes to an electric field that can be seen at an external point. A volumetric integration

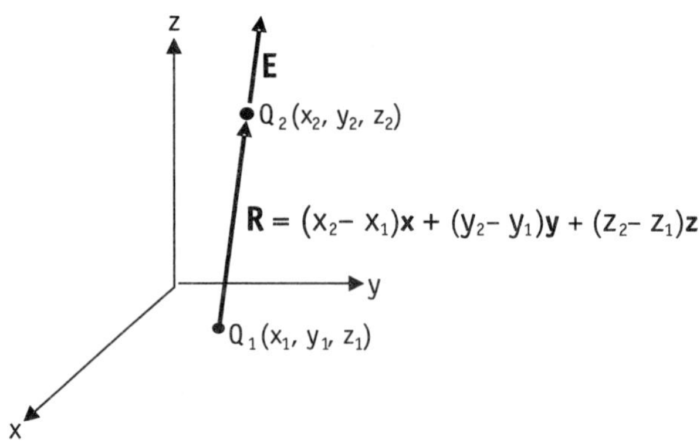

Figure 2.5 Illustration of electric field intensity vector (Cartesian coordinates).

is then required to obtain the total electric field. Even though electric charge in its smallest division is an electron or proton, it is useful to also consider charge as a continuous distribution, furthermore, one that is differentiable. As such, one may define and quantify the charge density by way of the following:

$$\rho_V = \frac{d}{dV}\ (Q) = \frac{dQ}{dV}\ (\text{C/m}^3) \tag{2.9}$$

The unit in parentheses is meant to signify that ρV will be in C/m^3, provided, of course, that the variables in the expression are quantified in proper SI units (i.e., C for charge, Q and m^3 for volume, V). This convention is used throughout this book.

Referring to the Figure 2.6 volume, V, each differential charge, dQ, produces a differential electric field (E-field) that is quantified as at the observation point P.

$$d\mathbf{E} = \left(\frac{dQ}{4\pi\varepsilon_0 R^2}\right)\mathbf{r}\ \ (\text{V/m}) \tag{2.10}$$

Under the assumption that the only charge in the region is contained within the volume, V, the total electric field at P is obtained by integration over the volume, that is:

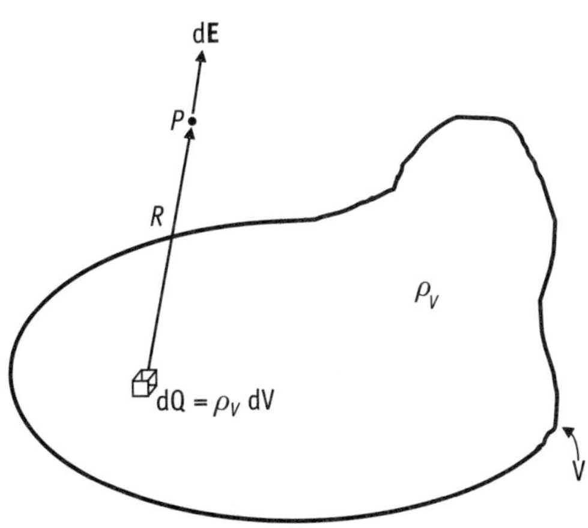

Figure 2.6 Generalized volumetric charge distribution.

$$\mathbf{E} = \int_V \left(\frac{\rho_V}{4\pi\varepsilon_0 R^2} \right) \mathbf{r}\, dV \quad \text{(V/m)} \tag{2.11}$$

A real-world application of a volumetric charge distribution is the case of commercial passenger or freighter airplane fuel tank being filled with Jet A-type kerosene. Before entering the fuel tank, the liquid dielectric (Jet A airplane fuel) passes through a hose that has an in-line particulate filter. While the filter removes impurities, that is, physical particulates, so that they do not enter the fuel tank, triboelectric friction created by the fuel passing across the filter elements generates electric charge. The generated electric charge flows down the remainder of the hose interior with the fuel to enter the tank through a designated opening. At the instant the fuel tank is filled with fuel, as indicated by the airplane fuel quantity sensors, like (positive) charge is distributed throughout the liquid fuel dielectric in a uniform charge distribution, making an easy-to-evaluate simplifying assumption.

Sheet charge. Charge may be also distributed over a surface or a sheet. Each differential charge, dQ, on the sheet results in a differential electric field,

$$d\mathbf{E} = \left(\frac{dQ}{4\pi\varepsilon_0 R^2} \right) \mathbf{r} \quad \text{(V/m)} \tag{2.12}$$

at point P (see Figure 2.7).

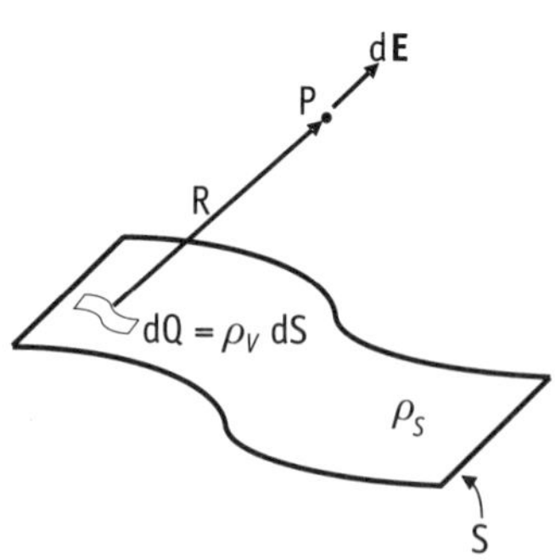

Figure 2.7 Generalized sheet (surface) charge distribution.

If the surface charge density is ρ_s (C/m^2) and no other charge is present in the region, then the total electric field at P is:

$$\mathbf{E} = \int_S \left(\frac{\rho_S}{4\pi\varepsilon_0 R^2} \right) \mathbf{r}\, dS \ \ (\text{V/m}) \tag{2.13}$$

Line charge. If charge is distributed over a line, each differential charge, dQ, along the sheet produces a differential electric field,

$$d\mathbf{E} = \left(\frac{dQ}{4\pi\varepsilon_0 R^2} \right) \mathbf{u}_R \ \ (\text{V/m}) \tag{2.14}$$

at point P (see Figure 2.8).

Furthermore, if the line charge density is ρ_l (C/m) and no other charge is in the region, the total electric field at P is,

$$\mathbf{E} = \int_L \left(\frac{\rho_l}{4\pi\varepsilon_0 R^2} \right) dl\ \mathbf{r} \ \ (\text{V/m}) \tag{2.15}$$

It should be emphasized that in all three of the above charge distributions and corresponding integrals for **E**, the unit variable, **r**, is variable depending on the coordinates of the charge element dQ. Thus, **r** cannot be removed from the integrand [1].

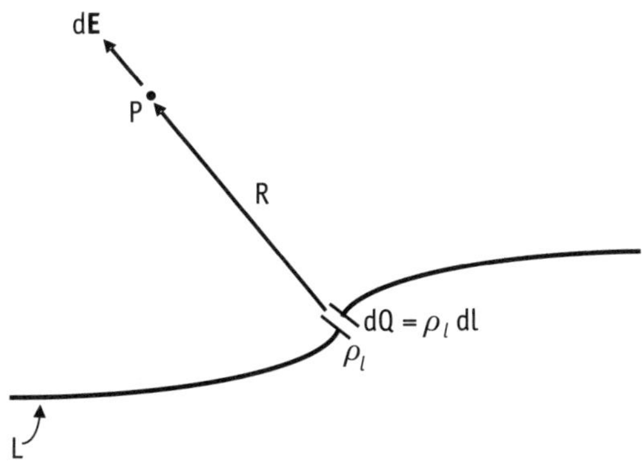

Figure 2.8 Generalized line charge distribution.

2.4 STANDARD CHARGE CONFIGURATIONS

In three special cases, the integration discussed in the previous section is either unnecessary or easily carried out. With regard to both these standard charge configurations (and to others that will be addressed later in the chapter), it should be pointed out that the charge is not on a conductor.

When a word problem states that charge is distributed in the form of a disk, for example, it does not mean that a disk-shaped conductor has charge on the surface. Although a stretch of the imagination may be required for this, these charges should be thought of as somehow suspended in space, fixed in the specified configuration.

Point charge. As determined in the previous section, the field of a single point charge Q is given by:

$$\mathbf{E} = \left(\frac{Q}{4\pi\varepsilon_0 R^2}\right)\mathbf{r} \text{ (spherical coordinates) (V/m)} \tag{2.16}$$

Figure 2.9 is a spherically symmetrical field that follows the inverse-square law (the same as for the force of gravitation).

Infinite line charge. If charge is distributed with uniform density, ρ_l (C/m) along an infinite straight line, which for convenience with be chosen as the z-axis (see Figure 2.10), then

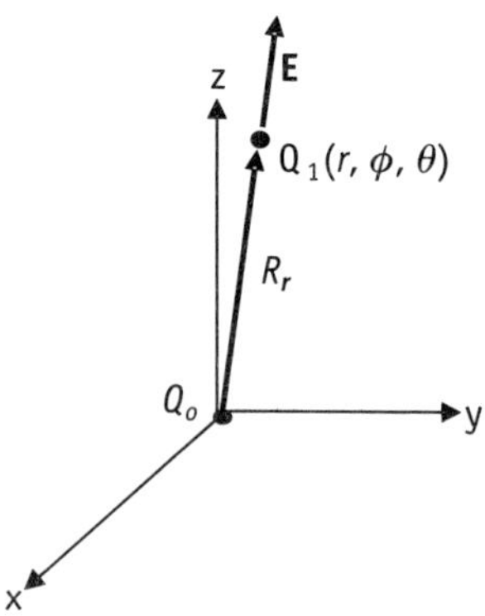

Figure 2.9 Spherically symmetric field (follows inverse-square law).

$$\mathbf{E} = \left(\frac{\rho_l}{2\pi\varepsilon_0 R} \right) \mathbf{r} \text{ (cylindrical coordinates) (V/m)} \tag{2.17}$$

This field has cylindrical symmetry and is inversely proportional to the first power of the distance from the infinite line charge.

Infinite plane charge. If charge is distributed with uniform density, ρ_S (C/m^2), over an infinite plane (see Figure 2.11), the field is then given by:

$$\mathbf{E} = \left(\frac{\rho_S}{2\varepsilon_0} \right) \mathbf{n} \tag{2.18}$$

This field has a constant magnitude and has mirror symmetry about the planar charge [2].

2.5 PROBLEMS: COULOMB FORCE AND E-FIELD INTENSITY

Problem 2.1

Find the force, $\mathbf{F}_1$, on Q_1 located at point (0, 1, 2) with charge of 20 μC due to charge Q_2 located at point (2, 0, 0) having charge of −300 μC.

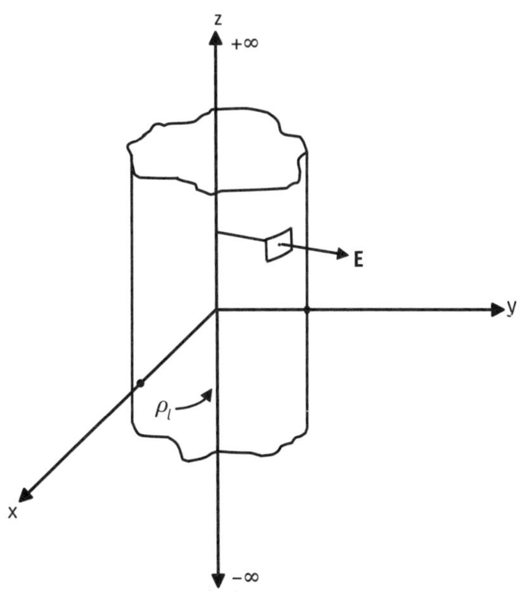

Figure 2.10 Infinite line charge, cylindrically symmetrical field.

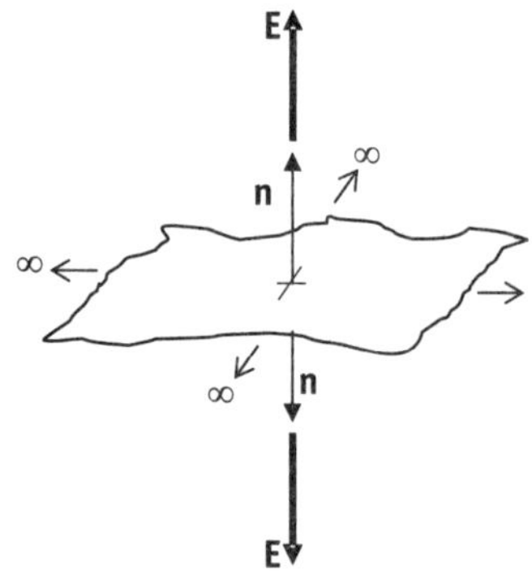

Figure 2.11 Infinite plane charge.

Problem 2.2

Two point charges, Q_1 and Q_2, are located at two separate points in space. Write a generalized equation to find the force of Q_2 on Q_1.

Problem 2.3

Two point charges, $Q_1 = 50\ \mu C$ and $Q_2 = 10\ \mu C$, are located, respectively, at (−1, 1, −3) and (3, 1, 0). Find the force on Q_1.

Problem 2.4

A point charge of $Q_1 = 300\ \mu C$ at (1, −1, −3)m experiences a force, $\mathbf{F} = 8\mathbf{x} - 8\mathbf{y} + 4\mathbf{z}$ N due to point charge Q_2 at (3, −3, −2). Determine Q_2.

Problem 2.5

Find the force on a point charge of 50 μC located at (0, 0, 5)m due a charge of $500\pi\ \mu C$ uniformly distributed over the circular disk $r \leq 5$m, $z = 0$m.

Problem 2.6

Develop an expression for **E** due to charge uniformly distributed over an infinite plane with density, ρ_s.

Problem 2.7

Charge is distributed uniformly along an infinite straight line with density ρ_l. Develop the expression for **E** at the general point P.

Problem 2.8

Find the expression for **E** at the point P due to the point charge Q located at (x_1, y_1, z_1).

Problem 2.9

Find **E** at the point P (0, 0, 5) due to the point charge $Q_1 = 0.35\ \mu\text{C}$ located at (0, 4, 0) and $Q_2 = -0.55\ \mu\text{C}$ located at (3, 0, 0).

Problem 2.10

On the line described by $x = 2\text{m}$, $y = -4\text{m}$, that is, $(2, -4, z)$, there is a uniform charge distribution of density $\rho_l = 20\ \text{nC/m}$. Describe the field, **E**, at location (−2, −1, 4).

Problem 2.11

Determine **E** at (2, 0, 2)m due to three standard charge distributions as follows:

- A uniform sheet a $x = 0\text{m}$ with $\rho_{S1} = (1/3\pi)\ \text{nC/m}^2$;
- A uniform sheet a $x = 4\text{m}$ with $\rho_{S2} = (-1/3\pi)\ \text{nC/m}^2$;
- A uniform sheet a $x = 4\text{m}$, $y = 0\text{m}$ with $\rho_l = -2\ \text{nC/m}$.

Problem 2.12

Charge is distributed along the z-axis between $z = \pm 5\text{m}$ with uniform density $\rho_l = 20\ \mu\text{C/m}$. Determine **E** at (2, 0, 0) in:

(a) Cartesian coordinates;

(b) Cylindrical coordinates.

Problem 2.13

Find the electric field intensity, **E**, at the point $P(0, \phi, h)$ in cylindrical coordinates due to the uniformly charged disk $r \leq a$, $z = 0$.

Problem 2.14

Charge lies on a circular disk $r \leq a$ and $z = 0$ that has a charge density, $\rho_S = \rho_o \sin^2\phi$. Determine **E** at $(0, \phi, h)$.

Problem 2.15

Charge lies on the circular disk $r \leq 4$m, $z = 0$ with uniform density $\rho_S = 10^{-4}/r$ (C/m^2). Determine **E** at $r = 0$, $z = 3$m.

Problem 2.16

Charge line in the $z = -3$m plane in the form of a square sheet as defined by

$$2\text{m} \leq x \leq -2\text{m}$$

and

$$2\text{m} \leq y \leq -2\text{m}$$

with charge density $\rho_S = 2(x^2 + y^2 + 9)^{3/2}$ (nC/m^2). Find **E** at the origin.

Problem 2.17

A charge of uniform density, $\rho_S = 0.3$ nC/m^2 covers the plane of $2x - 3y + z = 6$m. Find **E** on the side of the plane facing the origin.

Problem 2.18

Two point charges, $Q_1 = 250\ \mu$C and $Q_2 = -300\ \mu$C, are located at (5, 0, 0)m and (0, 0, −5)m, respectively. Find the force on Q_2.

Problem 2.19

Two point charges, $Q_1 = 30\ \mu$C and $Q_2 = -100\ \mu$C, are located at (2, 0, 5)m and (−1, 0, −2)m, respectively. Find the force on Q_1.

References

[1] Shadowitz, A., *The Electromagnetic Field*, New York: Dover Publications, Inc., 1975.

[2] Russer, P., *Electromagnetics, Microwave Circuit and Antenna Design for Communications Engineering*, Norwood, MA: Artech House, 2006.

[3] Warnick, K. F., and P. Russer, *Problem Solving in Electromagnetics, Microwave Circuit and Antenna Design for Communications Engineering*, Norwood, MA: Artech House, 2006.

[4] Cheng, D. K., *Fundamentals of Engineering Electromagnetics*, Reading, MA: Addison-Wesley Publishing Company, Inc., 1993.

3

ELECTRIC FLUX AND GAUSS' LAW

In the previous chapter, Coulomb's law was introduced to the reader. Along with it were the electrostatic forces of repulsion (between like-charges) and attraction (between unlike charges). Also introduced was the concept of electric field intensity, and with it the concepts of volume charge, infinite sheet charge, infinite line charge, and point charge.

In this chapter, determination of net charge in a volume will be described along with the concepts of electric flux and flux density. The origin and termination points of these are illustrated and described. Gauss' law is introduced along with the relation between electric flux and flux density. Finally, the important and especially helpful special Gaussian surfaces (i.e., the imaginary Gaussian spherical shell (sphere) and special Gaussian cylinder are introduced).

3.1 NET CHARGE IN A REGION

With charge density defined in Section 2.3, it is at this point possible to quantify the net charge contained in a specified volume by way of integration. From:

$$dQ = \rho_V dV \quad \text{(C)} \tag{3.1}$$

it follows that:

$$Q = \int_V \rho_V \, dV \quad \text{(C)} \tag{3.2}$$

Of course, ρ_V need not be constant throughout the volume, V [1].

3.2 ELECTRIC FLUX AND FLUX DENSITY

By definition and convention, electric flux, Ψ, originates on positive charge and terminates on negative charge. In the absence of a terminating negative charge, the flux, Ψ, terminates at infinity.

Also, by definition, one coulomb (C) of electric charge gives rise to one coulomb of electric flux. Hence:

$$\psi = Q \quad \text{(C)} \tag{3.3}$$

In Figure 3.1, the flux lines leave $+Q$ and terminate on $-Q$. The assumption in the figure is that the two unlike charges are of equal magnitude.

The case of the positive charge with no negative charge to terminate the flux, as shown in Figure 3.2, illustrates that flux lines are equally spaced throughout the solid angle and reach out to infinity.

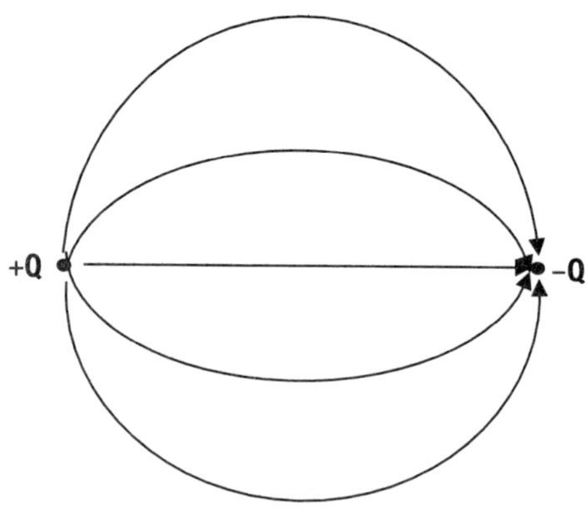

Figure 3.1 Origination and termination of electric flux, point charge to point charge.

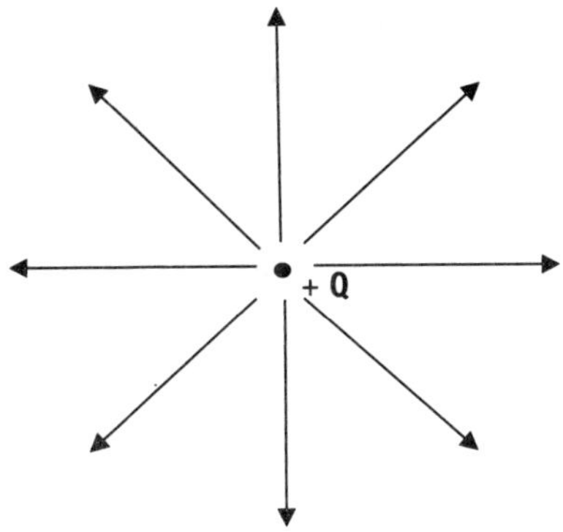

Figure 3.2 Origination and termination of electric flux point to infinity in all directions.

While the electric flux, Ψ, is a scalar quantity, the *density of electric flux*, **D**, is a vector field. As such, it takes its direction from the lines of flux. If in the neighborhood of point P, the flux lines have the direction of the unit vector, **a**, as illustrated in Figure 3.3.

Furthermore, if an amount of flux, $d\psi$, crosses the differential area, dS, which is normal to **a**, then the electric flux density at P is [2]:

$$\mathbf{D} = \frac{d\Psi}{dS}\mathbf{a} \quad \left(\mathrm{C/m^2}\right) \tag{3.4}$$

A volume charge distribution of density, ρ_V (C/m³), is shown enclosed by surface, S, in Figure 3.4.

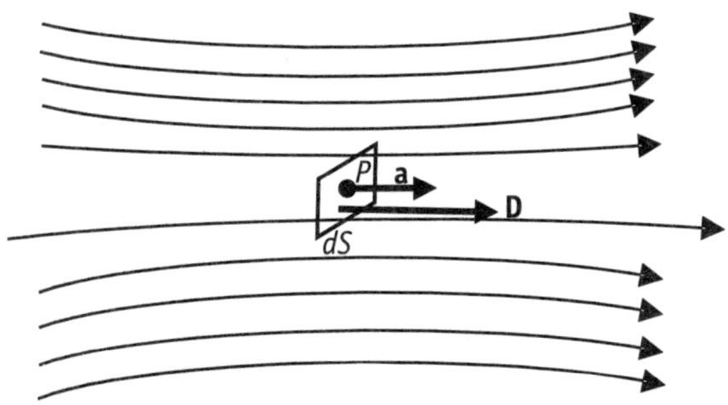

Figure 3.3 Lines of flux through point P and an elemental surface area, dS.

Since each coulomb of charge Q has, by definition, 1C of flux, Ψ, it follows that the net flux crossing the closed surface, S, is an exact measure of net charge enclosed. This said, the density, $\mathbf{D}$, may vary in magnitude and direction from point to point on S. So, generally speaking, $\mathbf{D}$ will not be along the normal line to S. If at the surface of element dS, $\mathbf{D}$ makes an angle θ with the normal line, then the differential flux crossing dS is given by

$$d\Psi = DdS\cos\theta = \mathbf{D}\cdot dS\,\mathbf{a}_n = \mathbf{D}\cdot d\mathbf{S} \tag{3.5}$$

where $d\mathbf{S}$ = the vector surface element of magnitude dS and direction $\mathbf{a}_n$.

The unit vector, $\mathbf{a}_n$ (a.k.a., $\mathbf{u}_n$ or $\mathbf{n}$) is always taken to point out of S so that $d\Psi$ is the amount of flux passing from the interior of S to the exterior of S through dS [3].

3.3 GAUSS' LAW

Integration of the above expression for $d\Psi$ over the closed surface S gives since $\Psi = Q$,

$$\oint \mathbf{D}\cdot d\mathbf{S} = Q_{\text{enc}} \text{ (Gauss' Law)} \tag{3.6}$$

The above equation is Gauss' law. It states that the total flux out of a closed surface is equal to the net charge within that surface.

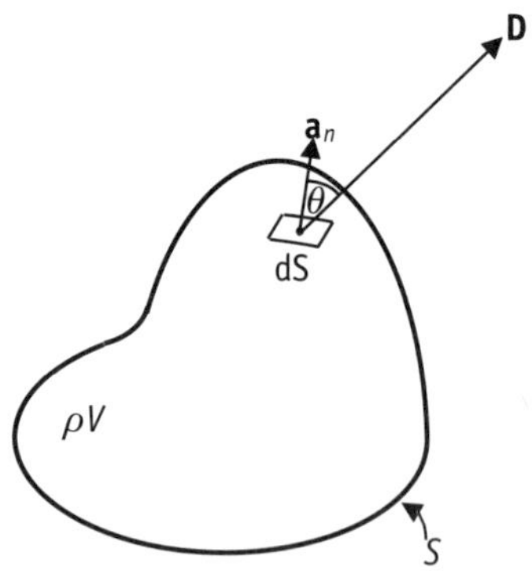

Figure 3.4 A volume charge density, ρ_V (C/m^3) with a surface area enclosing the volume.

It will be seen that a great deal of valuable information can be obtained from applying Gauss' law without having to perform the integration [2].

3.4 RELATION BETWEEN FLUX DENSITY AND ELECTRIC FIELD INTENSITY

Consider a point charge Q—assumed to be positive for simplicity—that is located at the origin (see Figure 3.5).

If the point charge is enclosed by a spherical surface having a radius, r, then by symmetry, **D**, due to Q, is of constant magnitude over the surface. What's more, it is *everywhere* normal to the surface. Gauss' law then gives:

$$Q = \oint \mathbf{D} \cdot d\mathbf{S} = D\oint dS = D\left(4\pi r^2\right) \tag{3.7}$$

from which $D = Q/4\pi r^2$. For this reason,

$$\mathbf{D} = \frac{Q}{4\pi r^2}\mathbf{a}_n = \frac{Q}{4\pi r^2}\mathbf{a}_r = \frac{Q}{4\pi r^2}\mathbf{n} = \frac{Q}{4\pi r^2}\mathbf{r} \tag{3.8}$$

However, from Section 2.2, the electric field intensity due to Q is:

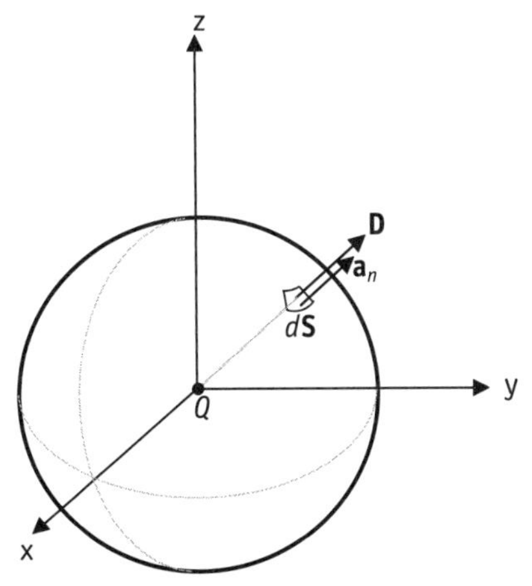

Figure 3.5 Point charge, Q, enclosed by a sphere (Gauss' law illustration).

$$\mathbf{E} = \frac{Q}{4\pi\varepsilon_0 r^2}\mathbf{a}_r = \frac{Q}{4\pi\varepsilon_0 r^2}\mathbf{r} \tag{3.9}$$

It follows that:

$$\mathbf{D} = \varepsilon_0 \mathbf{E} \tag{3.10}$$

More generally speaking, any electric field in an isotropic medium of permittivity, ε, will have

$$\mathbf{D} = \varepsilon \mathbf{E} \tag{3.11}$$

Thus, **D** and **E** fields will have identically the same form. They differ only by a factor which is a constant of the dielectric medium involved. The *e*-field, **E**, due to a charge configuration is a function of the permittivity ε. However, the electric flux density **D** is not. In situations involving multiple dielectric materials, a distinct advantage will be found in first obtaining **D** and then later converting to **E** for each dielectric material [2, 3].

3.5 SPECIAL GAUSSIAN SURFACES

The spherical surface used in Section 3.4 was a special Gaussian surface. That Gaussian surface has, and satisfies, the following defining conditions:

1. The surface is closed;
2. **D** is either normal or tangential to the surface at each point of the surface;
3. D has the same value at all points of the surface where **D** is normal.

It is a special Gaussian surface known as the spherical surface (see Figure 3.6).

A second type of special Gaussian surface is the cylindrical surface (see Figure 3.7).

One serious limitation to the method of special Gaussian surfaces is they can only be utilized for *highly symmetrical* charge

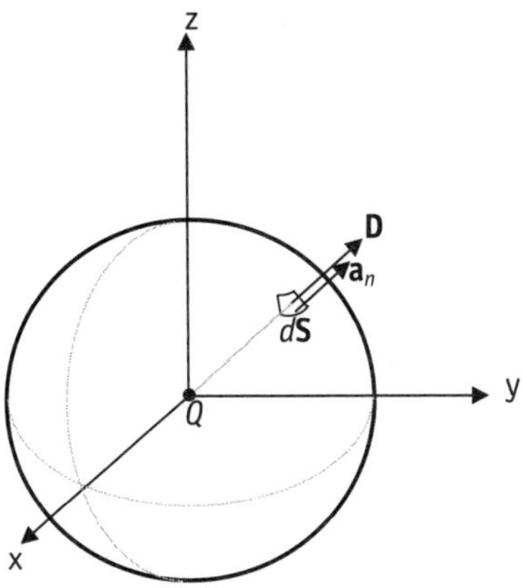

Figure 3.6 Special Gaussian surface: spherical surface.

configurations. For other configurations, the method can yet provide quick approximations to the field at locations very close to or very far from the charges [1].

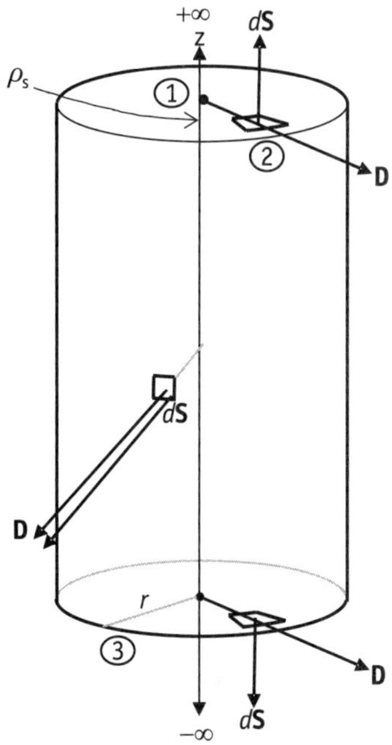

Figure 3.7 Special Gaussian surface: cylindrical surface.

3.6 PROBLEMS AND SOLUTIONS: ELECTRIC FLUX AND GAUSS' LAW

It is important to note the following:

1. **D** and **E** fields will have exactly the same form, differing only by ε, constant of the medium.
2. In problems involving multiple dielectrics, a distinct advantage will be found in first obtaining **D** and then converting to **E** within each dielectric.
3. The special Gaussian surfaces are conceptual geometric shapes (i.e., the imaginary spherical shell and the cylindrical shell), that help to more easily find solutions to electrostatics problems [3].

Problem 3.1

Use the special Gaussian surface to find D due to a uniform line charge, ρ_l (C/m).

Problem 3.2

Use the special Gaussian surface, the imaginary sphere shell, to find **D** due to point charge, Q (C), located at the sphere's center.

Problem 3.3

Find **E** in Problem 3.2.

Problem 3.4

Find **E** in Problem 3.1.

Problem 3.5

Find the charge in the volume shown in Figure 3.6 defined by:

(a) $0 \leq x \leq 1\text{m}$
$0 \leq y \leq 1\text{m}$
$0 \leq z \leq 1\text{m}$

and

(b) $0 \leq x \leq 1\text{m}$
$-1 \leq y \leq 0$
$0 \leq z \leq 1\text{m}$

if $\rho_V = 30\, x^2 y$ (μC/m^3).

Problem 3.6

Find the charge in the volume defined by $1 \leq r \leq 2$m in spherical coordinates if $\rho_V = 5 \cos^2\phi/r^4$ (C/m^3).

Problem 3.7

Three point charges (i.e., $Q_1 = 30$ nC, $Q_2 = 150$ nC, and $Q_3 = -70$ nC), are enclosed by a surface, S. What net flux crosses S?

Problem 3.8

What net flux crosses the closed surface, S, which contains a charge distribution in the form of a plane disk of radius 4m having a density of $\rho_S = \sin\phi$ (μC/m^2)?

Problem 3.9

A circular disk of radius 4m having a surface charge density of $\rho_S = 12\sin\phi \times 10^{-6}$ C/m^2 is enclosed by surface S. What net flux crosses S?

Problem 3.10

Two charges of the same magnitude but opposite signs are enclosed by surface, S. Can flux Ψ cross the surface?

Problem 3.11

Charge in the form of a plane sheet with density $\rho_S = 40$ μC/m^2 is located at $z = 0.5$m. A uniform charge line of $\rho_l = -6$ μC/m lies along the y-axis. What net flux crosses the surface of a cube 2m on an edge centered at the origin?

Problem 3.12

A point charge Q is at the origin of a spherical coordinate system.

(a) Find the flux that crosses the portion of a spherical shell described by $\alpha \leq \phi \leq \beta$.
(b) What is the result if $\alpha = 0$ and $\beta = \pi/2$?

Problem 3.13

A uniform line charge with $\rho_l = 50$ μC/m lies along the x-axis.

What flux per unit length, Ψ/L, crosses the portion of the $z = -3$m plane bounded by $y = \pm 2$m?

Problem 3.14

A point charge, $Q = 30$ nC, is located at the origin in Cartesian coordinates. Find the electric flux density, **D**, at (1, 3, −4)m.

Problem 3.15

Two identical uniform line charges lie along the x- and y-axes with charge densities, $\rho_l = 20\ \mu$C/m. Obtain **D** at (3, 3, 3)m.

Problem 3.16

Determine the flux crossing an area of 1 square millimeter on the surface of a cylinder shell at:

$R = 10$m, $\Phi = 53.2°$, and $Z = 2$m

if the surface area flux density (C/m^2) is:

$\mathbf{D} = (2x)\mathbf{x} + 2(1 - y)\mathbf{y} + (4z)\mathbf{z}$ (C/m^2).

Problem 3.17

Given an electric flux density, $\mathbf{D} = (2x)\mathbf{x} + 3\mathbf{y}$ (C/m^2), determine the net flux crossing the surface of a cube 2m on an edge centered at the origin (cube edges are parallel to the Cartesian coordinate axes).

Problem 3.18

A uniform line charge of $\rho_l = 3\ \mu$C/m lies along the z-axis, and a concentric cylinder of $r = 2$m has a surface charge distribution of $\rho_S = -(1.5/4\pi)\ \mu$C/m^2. Both distributions are infinite in extent with the z-axis. Use Gauss' law to find **D** in region.

Problem 3.19

A charge configuration in cylindrical coordinates is given by $\rho_V = 5re^{-2r}$ (C/m^3).

References

[1] Shadowitz, A., *The Electromagnetic Field*, New York: Dover Publications, Inc., 1975.

[2] Cheng, D. K., *Fundamentals of Engineering Electromagnetics*, Reading, MA: Addison-Wesley Publishing Company, Inc., 1993.

[3] Russer, P., *Electromagnetics, Microwave Circuit and Antenna Design for Communications Engineering*, Norwood, MA: Artech House, 2006.

[4] Warnick, K. F., and P. Russer, *Problem Solving in Electromagnetics, Microwave Circuit and Antenna Design for Communications Engineering*, Norwood, MA: Artech House, 2006.

4

DIVERGENCE AND THE DIVERGENCE THEOREM

In this chapter, an introductory discussion on divergence is undertaken, and it is undertaken in the three coordinate sytem forms: (1) Cartesian, (2) cylindrical, and (3) spherical. The del operator is introduced as is divergence theorem. Problems and solutions involving all the above are given at the end of the chapter.

4.1 DIVERGENCE

There are two characteristics regarding the manner in which a vector field changes from point to point in space. The first of these characteristics is *divergence*. Divergence is a *scalar* quantity, bearing similarity to the derivative of a function. The second characteristic is *curl*, a vector to be examined during the discussion of magnetic fields in Volume 2 [1].

When the divergence of a vector field is nonzero, the region is said to contain *sources* or *sinks*. Sources occur when the divergence is positive. Sinks occur when the divergence is negative.

In static electric fields, a correspondence exists between positive divergence, sources, and positive electric charge Q. Electric flux, Ψ, by definition, originates on positive charge. Therefore, a region that contains positive charges also contains the sources of Ψ. Divergence of the electric flux density $\mathbf{D}$ will be positive in this region. A similar correspondence exists as well between negative divergence, sinks, and negative electric charge.

Divergence of the vector field **A** at the point P is defined by:

$$\text{div}A = \lim_{\Delta V \to 0} \oint \mathbf{A} \cdot \frac{d\mathbf{S}}{\Delta V} \tag{4.1}$$

where V = volume.

In this case, the integration is over the surface of an infinitesimal volume, ΔV, which shrinks to point P.

4.2 DIVERGENCE IN CARTESIAN COORDINATES

The divergence can be expressed for any vector field and in any coordinate system. For development in Cartesian coordinates, a cube is selected with edges Δx, Δy, and Δz parallel to the x-, y-, and z-axes as illustrated in Figure 4.1.

The vector field, **A**,

$$\mathbf{A} = A_x\mathbf{x} + A_y\mathbf{y} + A_z\mathbf{z} \tag{4.2}$$

is then defined at point P; that is, at the cube corner with the lowest values of x, y, and z coordinates.

In order to express $\oint \mathbf{A} \cdot d\mathbf{S}$ for the cube, all six faces must be addressed in the analysis. On each face, the direction of $d\mathbf{S}$ is out-

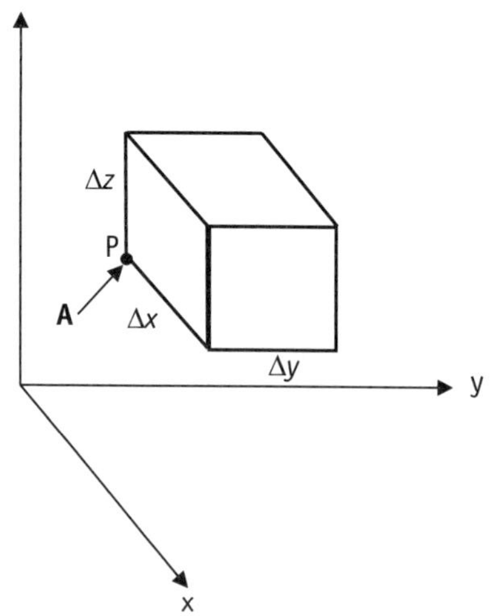

Figure 4.1 Cartesian coordinate divergence illustration.

ward. Since the faces are normal to the three coordinate axes, only one component of **A** crosses any two parallel faces (see Figure 4.2).

In Figure 4.2, the cube is turned such that face 1 is in full view. The x-components of A that are over the faces to the left and right of face 1 are indicated. Since the faces are small [2],

Left face:

$$\int \mathbf{A} \cdot d\mathbf{S} \sim -A_x(x)\Delta y\Delta z \tag{4.3}$$

Right face:

$$\int \mathbf{A} \cdot d\mathbf{S} \sim A_x(x+\Delta x)\Delta y\,\Delta z = \left[A_x(x) + \left(\frac{\partial A_x}{\partial x}\right)\Delta x\right]\Delta y\,\Delta z \tag{4.4}$$

the total for these two faces is:

$$\left(\frac{\partial A_x}{\partial x}\right)\Delta x\,\Delta y\,\Delta z \tag{4.5}$$

The same procedure is applied to the remaining two pairs of faces and the results combined.

$$\int \mathbf{A} \cdot d\mathbf{S} = \left(\frac{\partial A_x}{\partial x} + \frac{\partial A_y}{\partial y} + \frac{\partial A_z}{\partial z}\right)\Delta x\,\Delta y\,\Delta z \tag{4.6}$$

Dividing by

$$\Delta x\,\Delta y\,\Delta z = \Delta V \tag{4.7}$$

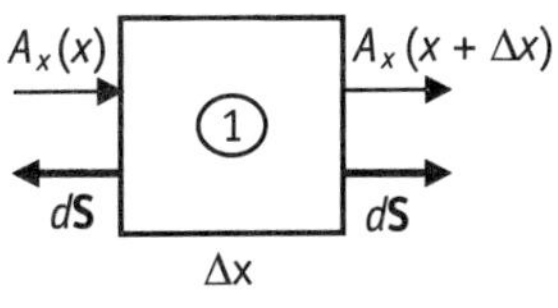

Figure 4.2 Cartesian coordinate divergence illustration, cube turned, face in full view.

and letting $\Delta V \rightarrow 0$, one obtains:

$$\operatorname{div}\mathbf{A} = \left(\frac{\partial A_x}{\partial x} + \frac{\partial A_y}{\partial y} + \frac{\partial A_z}{\partial z}\right) \text{ (expression in Cartesian coordinates)} \tag{4.8}$$

The same approach may be used in the use of cylindrical and spherical coordinates:

Cylindrical coordinates:

$$\operatorname{div}\mathbf{A} = \left[\left(\frac{1}{r}\right)\frac{\partial}{\partial r}\left(rA_r\right) + \left(\frac{1}{r}\right)\frac{\partial A_\phi}{\partial \phi} + \frac{\partial A_z}{\partial z}\right) \tag{4.9}$$

Spherical coordinates:

$$\operatorname{div}\mathbf{A} = \left\{\left(\frac{1}{r^2}\right)\frac{\partial}{\partial r}\left(r^2 A_r\right) = \frac{1}{(r\sin\theta)}\left[\frac{\partial}{\partial \theta}\left(A_\theta \sin\theta\right)\right] + \left(\frac{1}{\sin\theta}\right)\left(\frac{\partial A_\phi}{\partial \phi}\right)\right\} \tag{4.10}$$

4.3 DIVERGENCE OF *D*

From Gauss' law (as described in Section 3.3),

$$\frac{\left(\oint \mathbf{D}\cdot d\mathbf{S}\right)}{\Delta V} = \frac{Q_{\text{enc}}}{\Delta V} \tag{4.11}$$

In the limit,

$$\lim_{\Delta V \to 0} \oint \mathbf{D}\cdot\frac{dS}{\Delta V} = \operatorname{div}\mathbf{D} = \lim_{\Delta V \to 0}\left(\frac{Q_{\text{enc}}}{\Delta V}\right) = \rho \tag{4.12}$$

This important result is one of James Clerk Maxwell's equations for static fields:

$$\operatorname{div}\mathbf{D} = \rho \tag{4.13}$$

and

$$\text{div}\,\mathbf{E} = \frac{\rho}{\varepsilon} \tag{4.14}$$

If ε is constant throughout the region under examination (if not, div $\varepsilon\,\mathbf{E} = \rho$). Thus, both **E** and **D** fields will have divergence of zero in any charge-free region.

4.4 THE DEL OPERATOR

Vector analysis has its own shorthand, and, as such, the reader must note it with care. At this point in the text, a vector operator, symbolized by ∇ is defined—in Cartesian coordinates—by:

$$\nabla = \frac{\partial(\;)}{\partial x}\mathbf{x} + \frac{\partial(\;)}{\partial y}\mathbf{y} + \frac{\partial(\;)}{\partial z}\mathbf{z} \tag{4.15}$$

Please note that the "( )" indicates the place in (4.15) on which to perform the del operator. In calculus, a differential operator D is sometimes used to represent d/dx. Note that the symbol's square root radical and $\int$ are also operators. Standing alone without any indication of what they are to operate on, they look strange [3].

Likewise, ∇ standing alone, simply suggests the taking of certain partial derivatives, each being followed by a unit vector. However, when ∇ is dotted with vector **A**, the result is the divergence of **A**.

$$\begin{aligned}\nabla \cdot \mathbf{A} &= \left(\frac{\partial}{\partial x}\mathbf{a}_x + \frac{\partial}{\partial y}\mathbf{a}_y + \frac{\partial}{\partial z}\mathbf{a}_z\right)\cdot\left(A_x\mathbf{a}_x + A_y\mathbf{a}_y + A_z\mathbf{a}_z\right)\\ &= \frac{\partial}{\partial x}\left(A_x\right) + \frac{\partial}{\partial y}\left(A_y\right) + \frac{\partial}{\partial z}\left(A_z\right) = div\,\mathbf{A}\end{aligned} \tag{4.16}$$

From here and forward, the divergence of a vector field, **A**, will be written as $\nabla \cdot \mathbf{A}$.

WARNING! *The del operator is defined ONLY in Cartesian coordinates.* When $\nabla \cdot \mathbf{A}$ is written for the divergence in other coordinate systems, it does not mean that a del operator can be defined for these systems. For example, the divergence in cylindrical coordinates will be written as:

$$\nabla \cdot \mathbf{A} = \left(\frac{1}{r}\right)\frac{\partial}{\partial r}\left(rA_r\right) + \left(\frac{1}{r}\right)\frac{\partial}{\partial \phi}\left(A_\phi\right) + \frac{\partial}{\partial z}\left(A_z\right) \quad \text{(See Section 4.2)} \tag{4.17}$$

It should be noted that *this does not imply* that:

$$\nabla = \left(\frac{1}{r}\right)\frac{\partial}{\partial r}(\ )\mathbf{a}_r + \left(\frac{1}{r}\right)\frac{\partial}{\partial \phi}(\ )\mathbf{a}_\phi + \frac{\partial}{\partial z}(\ )\mathbf{a}_z \tag{4.18}$$

in cylindrical coordinates. In fact, the expression would give false results when used in ∇V; that is, the gradient (see Chapter 5) or $\nabla \times \mathbf{A}$ (i.e., the curl). (See Volume 2 of this book.)

4.5 DIVERGENCE THEOREM

Gauss' law states that the enclosed surface integral of $\mathbf{D} \cdot d\mathbf{S}$ is equal to the enclosed charge if the charge density function, ρ_V, is known throughout the volume [4]. Thus,

$$\oint \mathbf{D} \cdot d\mathbf{S} = \int \rho_V \, dV = Q_{\text{enc}} \tag{4.19}$$

But,

$$\rho_V = \nabla \cdot \mathbf{D} \tag{4.20}$$

So,

$$\oint \mathbf{D} \cdot d\mathbf{S} = \int [\nabla \cdot \mathbf{D}]\, dV \tag{4.21}$$

This is the *divergence theorem* (also known as *Gauss' divergence theorem*); that is, a three-dimensional analog of Green's theorem for the plane. While it was derived from known relationships among **D**, Q, and ρ_V, the theorem is applicable to *any* vector field.

$$\text{Divergence theorem} \qquad \oint_S \mathbf{A} \cdot d\mathbf{S} = \int_V [\nabla \cdot \mathbf{A}]\, dV \tag{4.22}$$

Of course, the volume, V, is that which is enclosed by the surface, S.

The divergence theorem applies to time-variation fields as well as static fields in any coordinate system. The theorem is used most often in derivations where it becomes necessary to change from a closed surface integration to a volume integration. However, it may also be used in the reverse direction, too; that is, to convert the volume integral of a function that can be expressed as the divergence of a vector field into a closed surface integral.

4.6 PROBLEMS AND SOLUTIONS: DIVERGENCE AND THE DIVERGENCE THEOREM

Problem 4.1

Develop the expression for divergence in cylindrical coordinates (Figure 4.3).

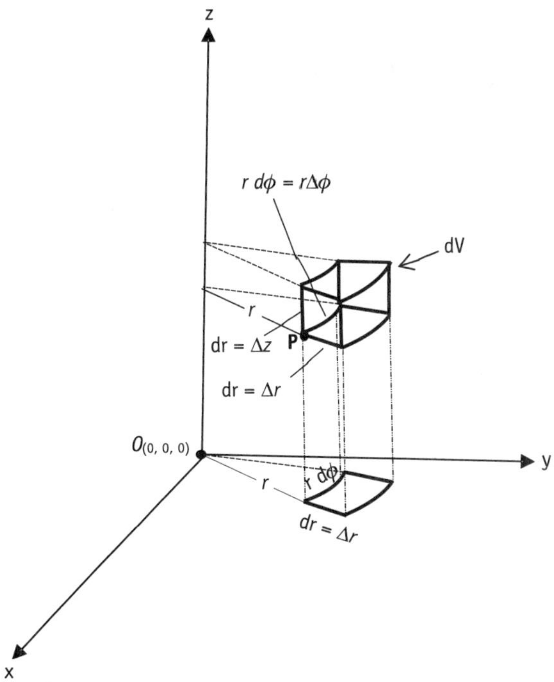

Figure 4.3 Illustration of delta volume with edges for Problem 4.1.

Problem 4.2

Show that $\nabla \cdot \mathbf{E} = 0$ for the field of a uniform line charge.

Problem 4.3

Show that the D field due to a point charge has a divergence of zero.

Problem 4.4

Given $\mathbf{A} = e^{-y}\,[\cos x\mathbf{x} - \sin x\mathbf{y}]$, find $\nabla \cdot \mathbf{A}$.

Problem 4.5

Given $A = x^2\mathbf{x} + (yz)\mathbf{y} + (xy)\mathbf{z}$, find $\nabla \cdot \mathbf{A}$.

References

[1] Cheng, D. K., *Fundamentals of Engineering Electromagnetics*, Reading, MA: Addison-Wesley Publishing Company, Inc., 1993.

[2] Shadowitz, A., *The Electromagnetic Field*, New York: Dover Publications, Inc., 1975.

[3] Russer, P., *Electromagnetics, Microwave Circuit and Antenna Design for Communications Engineering*, Norwood, MA: Artech House, 2006.

[4] Warnick, K. F., and P. Russer, *Problem Solving in Electromagnetics, Microwave Circuit and Antenna Design for Communications Engineering*, Norwood, MA: Artech House, 2006.

5

ENERGY AND ELECTRIC POTENTIAL OF CHARGE SYSTEMS

In this chapter, the concept of work is introduced as applicable to the movement of charge in free space. Discussion of electric potential between two points is undertaken as is an introduction to the gradient. The relationship between field intensity, **E**, and the gradient, ∇ is presented. The energy in static electric fields is also discussed and quantified.

5.1 WORK DONE IN MOVING A POINT CHARGE

In an electric field **E**, a point charge, Q, experiences a force that is given by

$$\mathbf{F} = Q\mathbf{E} \quad \text{(N)} \tag{5.1}$$

This unbalanced force will result in an acceleration of the charged particle with its motion in the direction of the field if Q is positive [1] (see Figure 5.1).

To put the charge in equilibrium, an *applied force* is required; that is, an applied force that is equal in magnitude and opposite in direction to the force from the field [1]:

$$F_a = -Q\mathbf{E} \quad \text{(N)} \tag{5.2}$$

Work is defined as a force acting over a distance [2]. Therefore, when the charge particle moves at constant speed through a differential distance, *dl*, the applied force does a differential amount of work, *dW*. The work may be positive or negative with respect to the applied force, $\mathbf{F}_a$, depending upon the direction of the vector displacement, $d\mathbf{l}$.

When $d\mathbf{l}$ and $\mathbf{F}_a$ are not in the same direction, the component of the force in the direction of $d\mathbf{l}$ must be used, expressed as:

$$dW = F_a dl \cos\theta = \mathbf{F_a} \cdot d\mathbf{l} \quad \text{(Joules)} \tag{5.3}$$

The differential work done by an external agent in an electric field is expressed as:

$$dW = -Q\mathbf{E} \cdot d\mathbf{l} \quad \text{(Joules)} \tag{5.4}$$

With this as the defining expression for work that is done in moving a charge particle in an electric field, a positive value means that work had to be done by an external agent in order to bring about the change in position. A negative result will mean that work was done by the field [3].

It is important to know the expressions for $d\mathbf{l}$ in the three coordinate systems [4]. These are:

$$d\mathbf{l} = dx\ \mathbf{x} + dy\ \mathbf{y} + dz\ \mathbf{z}; \quad \text{(Cartesian coordinates)} \tag{5.5}$$

$$d\mathbf{l} + dr\ \mathbf{r} + rd\phi\ \boldsymbol{\phi} + dz\ \mathbf{z}; \quad \text{(cylindrical coordinates)} \tag{5.6}$$

$$d\mathbf{l} = dr\ \mathbf{r} + r\ d\theta\ \boldsymbol{\theta} + r\sin\theta\ d\phi\ \boldsymbol{\phi}; \quad \text{(spherical coordinates)} \tag{5.7}$$

The work done in moving a point charge Q from point B to point A in a static electric field is the same for *any* path chosen. Equivalently, the work done in moving the charge around any closed loop is zero; that is,

$$\oint \mathbf{E} \cdot d\mathbf{l} = 0 \quad \text{(static fields)} \tag{5.8}$$

Such a vector field is called a *conservative* field [2].

5.2 ELECTRIC POTENTIAL BETWEEN TWO POINTS

The *potential* of point A with respect to point B is defined as the work done in moving a unit positive charge, Q_u, from point B to point A [1].

$$V_{\mathrm{AB}} = \frac{W}{Q_u} = -\int \mathbf{E} \cdot d\mathbf{l} \quad \text{(J/C or V)} \tag{5.9}$$

The terms potential and voltage are interchangeable.

It should be observed also that the initial, or reference, point is the lower limit of the line integral. Then, too, the minus sign must not be omitted. This sign came into the expression by way of the force $\mathbf{F}_a = -Q\mathbf{E}$ that had to be applied to put the charge in equilibrium.

Because **E** is a conservative field,

$$V_{\mathrm{AB}} = V_{\mathrm{AC}} - V_{\mathrm{BC}} \tag{5.10}$$

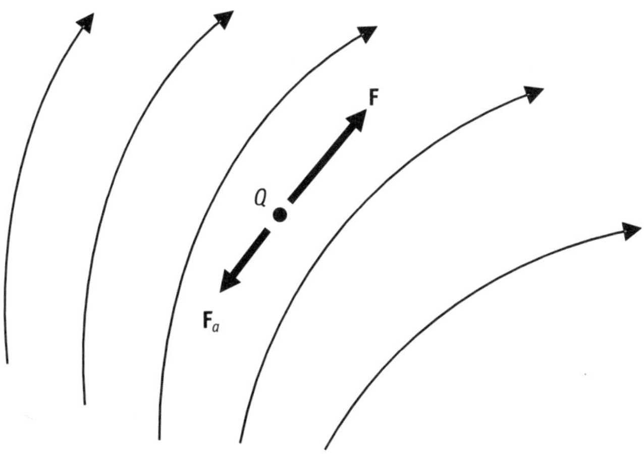

Figure 5.1 Force on a positive charge *Q* when moved from one point to another in an electric field.

V_{AB} may be considered now as the *potential difference* between points A and B. When V_{AB} is positive, work must be done to move the unit positive charge from point B to point A. Then, point A is said to be at a higher potential than point B. For example:

If a charge of +2C located at point B (2, 0, 0) on the graph in Figure 5.2, is moved to point A (0, 2, 0) and an energy of 24J is expended to do it, then

$$\begin{aligned} V_{AB} &= \frac{Q}{W} = \frac{24\text{J}}{+2\text{C}} \\ &= +12\text{V} \end{aligned} \tag{5.11}$$

where:

$Q = +2\text{C}$

$W = 24\text{J}$

It may be concluded then that point A is at a higher potential than B (i.e., 12V higher). Also, the potential V_{BA} must be −12V since V_{BA} differs from V_{AB} only by an interchange of the lower and upper limits in the defining integral, which simply changes the sign of the result.

5.3 POTENTIAL OF A POINT CHARGE

Since the electric field due to a point charge Q is completely in the radial direction, spherical coordinates are most conveniently used, and, moreover, only the radial (r-) component is needed [2]:

$$\begin{aligned} V_{AB} &= \int_{B}^{A} \mathbf{E} \cdot d\mathbf{l} = -\int_{r_B}^{r_A} E_r \, dr = -\left(\frac{Q}{4\pi\varepsilon_0}\right)\int_{r_B}^{r_A}\left(\frac{1}{r^2}\right)dr \\ &= \left(\frac{Q}{4\pi\varepsilon_0}\right)\left[\left(\frac{1}{r_A}\right)-\left(\frac{1}{r_B}\right)\right] \end{aligned} \tag{5.12}$$

For a positive charge Q, point A is at a higher potential than point B when r_A is smaller than r_B. The equipotential surfaces are concentric spherical shells.

If reference point B is now allowed to move out to infinity (∞), then:

$$V_{A\infty} = \left(\frac{Q}{4\pi\infty}\right)\left[\left(\frac{1}{r_A}\right) - \left(\frac{1}{\infty}\right)\right] \tag{5.13}$$

or

$$V = \left(\frac{Q}{4\pi\varepsilon_0 r}\right) \tag{5.14}$$

Considerable use will be made of this equation in the discussions that follow. The greatest danger to a problem-solution analyst is forgetting where the reference is while attempting to apply the equation to charge distributions which themselves extend to ∞. See Figure 5.2 for an illustration of work performed when a point charge is moved from one location to another.

5.4 POTENTIAL OF A CHARGE DISTRIBUTION

If the charge is distributed throughout some finite volume with a known charge density, ρ_V (unit: C/m^3), then the potential at some

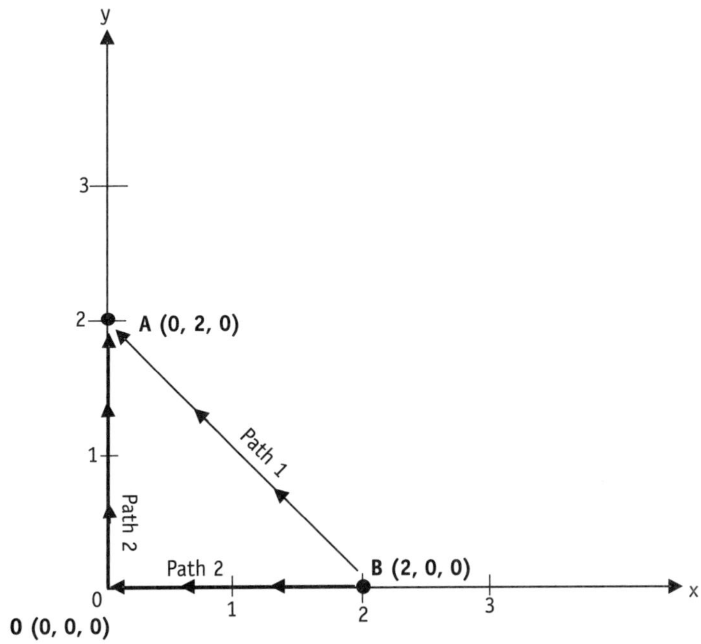

Figure 5.2 Illustration of work performed moving charge Q from one point to another in an electric field.

external point can be determined. To make this determination, a differential charge at an arbitrary point within the volume is identified (see Figure 5.3).

At point P:

$$dV = \frac{dQ}{4\pi\varepsilon_0 R} \tag{5.15}$$

Integration over the volume, *V*, gives the total potential (difference) at P.

Integral development:

$$dV = \frac{dQ}{4\pi\varepsilon_0 R} = \frac{(\rho_V\, dV)}{4\pi\varepsilon_0 R} \tag{5.16}$$

where $dQ = \rho_V\, dV$

$$\int_V dV = \int_V \frac{(\rho_V\, dV)}{4\pi\varepsilon_0 R} \tag{5.17}$$

to get:

$$V = \int \frac{(\rho_V\, dV)}{4\pi\varepsilon_0 R} = \int \left(\frac{\rho_V}{4\pi\varepsilon_0 R} \right) dV \tag{5.18}$$

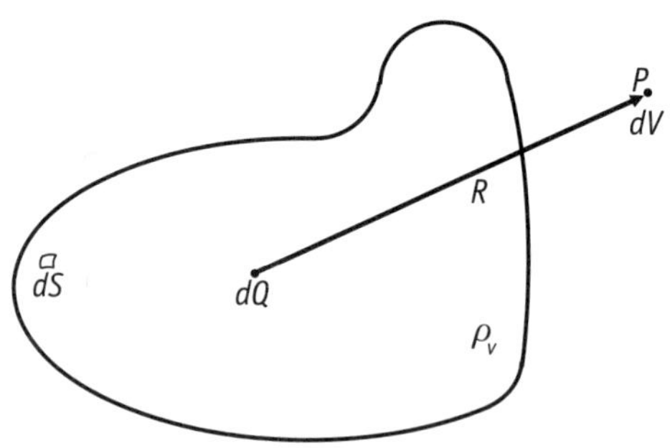

Figure 5.3 Arbitrary differential charge, *dQ*, in a volume of electric charge with charge density of ρ_V.

R must not be confused with r of the spherical coordinates system. It is not a vector but is the scalar distance from dQ to the fixed point, P. R always varies from place to place throughout the volume. As such, it cannot be removed from the integrand.

If charge is distributed over a surface or a line, the expression for V also holds, provided that integration is over the surface or the line with ρ_S or ρ_l used in place of ρ_V. It must be emphasized that all these expressions for the potential at an external point are based upon a *zero voltage (0V) reference at* ∞ [2].

5.5 GRADIENT

In this section, another vector analysis operation is introduced—*gradient*. Figure 5.4(a) shows two neighboring points of a region for which a scalar function is defined [3].

The vector separation of the two points is:

$$d\mathbf{r} = dx\mathbf{x} + dy\mathbf{y} + dz\mathbf{z} \tag{5.19}$$

From the calculus mathematics, the change in V from M to N is given by:

$$dV = \left(\frac{\partial V}{\partial x}\right)dx + \left(\frac{\partial V}{\partial y}\right)dy + \left(\frac{\partial V}{\partial x}\right)dz \tag{5.20}$$

Now the ∇ operator—introduced in Section 4.4—operating on V, gives:

$$\nabla V = \frac{\partial V}{\partial x}\mathbf{x} + \frac{\partial V}{\partial y}\mathbf{y} + \frac{\partial V}{\partial z}\mathbf{z} \tag{5.21}$$

It follows that:

$$dV = \nabla V \cdot d\mathbf{r} \tag{5.22}$$

The vector field ∇V—*also written as grad V*—is called the gradient of the scalar function V. It is seen that, for fixed $|d\mathbf{r}|$, the change in V in a given direction $d\mathbf{r}$ is proportional to the projection of ∇V in that direction. Thus, ∇V *lies in the direction of maximum increase of the function* V [2].

Another view of the gradient—see Figure 5.4(b)—is obtained by allowing the points M and N to lie on an equipotential surface, $V(x, y, z) = c_1$. Then $dV = 0$, which implies that ∇V is perpendicular to $d\mathbf{r}$. But $d\mathbf{r}$ is tangent to the equipotential surface. Indeed, for a suitable location of N, it represents *any* tangent through M. Therefore, ∇V must be along the surface normal at M. Since ∇V is in the direction of increasing V, it points from $V(x, y, z) = c_1$ to $V(x, y, z) = c_2$, where $c_2 > c_1$. Thus, the gradient of a potential function is a vector field that is everywhere normal to the equipotential surfaces.

The gradient in the cylindrical and spherical coordinate systems follows directly from that in the Cartesian system. It is noted

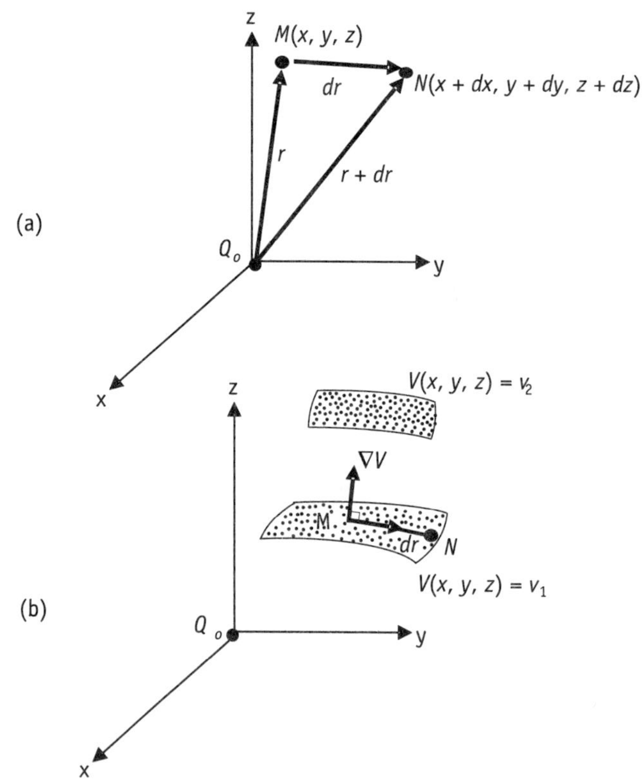

Figure 5.4 (a) Two neighboring points of a region in which a scalar function is defined, and (b) the same neighboring points on an equipotential surface.

that each term contains the partial derivative of V with respect to distance in the direction of that particular unit vector [4]:

$$\nabla V = \frac{\partial V}{\partial x}\mathbf{x} + \frac{\partial V}{\partial y}\mathbf{y} + \frac{\partial V}{\partial z}\mathbf{z} \quad \text{(Cartesian coordinates)} \tag{5.23}$$

$$\nabla V = \frac{\partial V}{\partial r}\mathbf{r} + \frac{\partial V}{r\,\partial \phi}\boldsymbol{\phi} + \frac{\partial V}{\partial z}\mathbf{z} \quad \text{(cylindrical coordinates)} \tag{5.24}$$

$$\nabla V = \frac{\partial V}{\partial r}\mathbf{r} + \frac{\partial V}{r\,\partial \theta}\boldsymbol{\theta} + \frac{\partial V}{[r \sin\theta\, \partial \phi]}\boldsymbol{\phi} \quad \text{(spherical coordinates)} \tag{5.25}$$

While ∇V is written for grad V in any coordinate system, it must be remembered that the ∇ operator is defined only in Cartesian coordinates [4].

5.6 RELATIONSHIP BETWEEN E AND ∇

From the integral expression for the potential of A with respect to B, the differential of V may be written

$$dV = -\mathbf{E} \cdot d\mathbf{l} \tag{5.26}$$

On the other hand,

$$dV = \nabla V \cdot d\mathbf{r} \tag{5.27}$$

Since $d\mathbf{l} = d\mathbf{r}$ is an arbitrary small displacement, it follows that:

$$\mathbf{E} = -\nabla V \tag{5.28}$$

The electric field intensity **E** may be obtained when the potential function V is known by simply taking the negative of the gradient of V. The gradient was found to be a vector normal to the equipotential surfaces, directed to a positive change in V. With the negative sign, the **E** field is found to be directed from higher to lower levels of the potential V [2].

5.7 ENERGY IN STATIC ELECTRIC FIELDS

Consider the work required to assemble, charge by charge, a distribution of $n = 3$ point charges. The region is assumed initially to be charge-free and with $\mathbf{E} = 0$ throughout. Referring to Figure 5.5, the work required to place the first charge, Q_1, into position 1 is zero.

Then, when Q_2 is moved toward the region, work equal to the product of this charge and the potential due to Q_1 is required. The total work to position the three charges is:

$$\begin{aligned} W_E &= W_1 + W_2 + W_3 \\ &= 0 + \left(Q_2 V_{2,1}\right) + \left(Q_3 V_{3,1} + Q_3 V_{3,2}\right) = W_{E1} \end{aligned} \tag{5.29}$$

The potential $V_{2,1}$ reads, "The potential at point 2 due to charge Q_1 at position 1." (Note: This unique notation will not appear again in the remainder of this book.) The work, W_E, is the energy stored in the electric field of the charge distribution.

If the three charges were brought into place in reverse order, the total work would be:

$$\begin{aligned} W_E &= W_3 + W_2 + W_3 \\ &= 0 + \left(Q_2 V_{2,3}\right) + \left(Q_1 V_{1,3} + Q_1 V_{1,2}\right) = W_{E2} \end{aligned} \tag{5.30}$$

When the two expressions, (5.29) and (5.30), are added, the result is twice the stored energy [2]:

$$\begin{aligned} 2W_E &= W_{E1} + W_{E2} \\ &= \left(W_1 + W_2 + W_3\right) + \left(W_3 + W_2 + W_1\right) \\ &= \left[0 + \left(Q_2 V_{2,1}\right) + \left(Q_3 V_{3,1} + Q_3 V_{3,2}\right)\right] \\ &\quad + \left[0 + \left(Q_2 V_{2,3}\right) + \left(Q_1 V_{1,3} + Q_1 V_{1,2}\right)\right] \\ &= Q_1\left(V_{2,1} + V_{1,3}\right) + Q_2\left(V_{2,1} + V_{2,3}\right) + Q_3\left(V_{3,1} + V_{3,2}\right) \end{aligned} \tag{5.31}$$

where

$W_{E1} = W_1 + W_2 + W_3;$
$W_{E2} = W_3 + W_2 + W_1;$
$W_1 + W_2 + W_3 = 0 + (Q_2V_{2,1}) + (Q_3V_{3,1} + Q_3V_{3,2});$
$W_3 + W_2 + W_1 = 0 + (Q_2V_{2,3}) + (Q_1V_{3,1} + Q_1V_{1,2}).$

The term $Q_1(V_{1,2} + V_{1,3})$ was the work done against the fields of Q_2 and Q_3, the only other charges in the region. Hence, $V_{1,2} + V_{1,3} = V_1$, the potential at position 1. Then,

$$\begin{aligned} 2W_E &= Q_1\left(V_{1,2} + V_{1,3}\right) + Q_2\left(V_{2,1} + V_{2,3}\right) + Q_3\left(V_{3,1} + V_{3,2}\right) \\ &= Q_1V_1 + Q_2V_2 + Q_3V_3 \end{aligned} \tag{5.32}$$

where

$V_1 = V_{1,2} + V_{1,3};$
$V_2 = V_{2,1} + V_{2,3};$
$V_3 = V_{3,1} + V_{3,2};$

and

$$W_E = \left(\frac{1}{2}\right) \sum_{m=3}^{n} Q_m V_m \tag{5.33}$$

for a region containing n point charges. For a region with a charge density, ρ_v (C/m^3), the summation becomes an integration [3]:

$$W_E = \left(\frac{1}{2}\right) \int \rho_v V \, dv \tag{5.34}$$

Other useful forms of the expression for stored energy are:

$$W_E = \left(\frac{1}{2}\right) \int \mathbf{D} \cdot \mathbf{E} \, dv \tag{5.35}$$

$$W_E = \left(\frac{1}{2}\right) \int \left(\varepsilon E^2\right) dv \tag{5.36}$$

and

$$W_E = \left(\frac{1}{2}\right)\int\left(\frac{D^2}{\varepsilon}\right)dv \tag{5.37}$$

In an electric circuit, the energy stored in the field of a capacitor is given by:

$$W_E = \left(\frac{1}{2}\right)QV = \left(\frac{1}{2}\right)CV^2 \tag{5.38}$$

where:

C is the capacitance (in Farads, F);

V is the voltage difference between the two conductors making up the capacitor (in volts, V);

Q is the magnitude of the total charge on one of the conductors (in coulombs, C).

5.8 PROBLEMS AND SOLUTIONS: ENERGY AND ELECTRIC POTENTIAL OF CHARGE SYSTEMS

Problem 5.1

Find the work done moving a charge of +2C from (2, 0, 0)m to (0, 2, 0)m along the straight line path joining the two points if the electric field is:

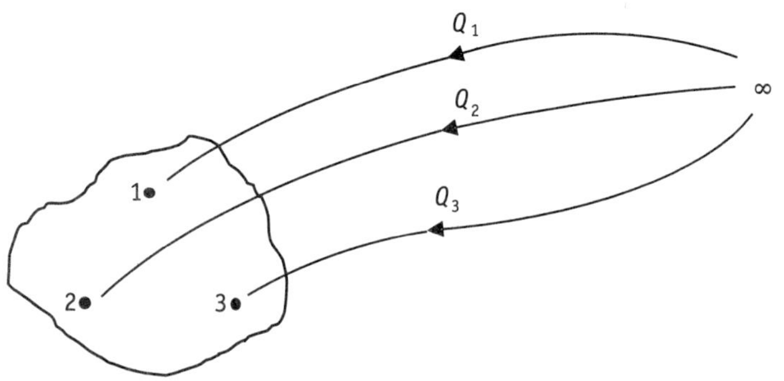

Figure 5.5 Work performed by moving charges Q_1, Q_2, and Q_3 from ∞ to points 1, 2, and 3.

$$\mathbf{E} = 2x\mathbf{x} - 4y\mathbf{y} \text{ (V/m)}$$

Problem 5.2

Find the work done in the field of the previous example problem when the +2C charge is moved from (2, 0, 0)m to (0, 0, 0)m along the x-axis and then from (0, 0, 0) to (0, 2, 0) along the y-axis.

Problem 5.3

Find the potential of A(1, ϕ, z) with respect to B(3, ϕ, z) in cylindrical coordinates where the electric field due to a line charge on the z-axis as given by $\mathbf{E} = 50\ (1/r)\mathbf{r}$ (V/m).

Problem 5.4

Find the work performed by an external agent applying a force to move a point charge of $Q = -20\ \mu$C from the origin to (4, 0, 0) in the field:

$$\mathbf{E} = (\tfrac{1}{2}x + 2y)\mathbf{x} + 2xy\mathbf{y} \text{ (V/m)}$$

Problem 5.5

In the field of Problem 5.1, move the charge from (4, 0, 0)m to (4, 2, 0)m and determine the work performed.

Problem 5.6

In the E-field of Problem 5.1, find the work performed by moving the charge from the origin to (4, 2, 0) along the straight line connecting the points.

Problem 5.7

Find the work done in moving a point charge of $Q = 5\ \mu$C from the origin to the spherical coordinate point (2m, $\pi/4$, $\pi/2$) in the field

$$\mathbf{E} = 5\ e^{-(1/4)r}\ \mathbf{r} + (10/r^2 \sin\theta)\ \boldsymbol{\varphi} \text{ (V/m).}$$

Problem 5.8

Given the field $\mathbf{E} = (k/r)\mathbf{r}$ in cylindrical coordinates, show that the work needed to move a point charge Q from any radial distance, r, to a point at twice that radial distance is independent of that distance.

Problem 5.9

For a line charge, ρ_l, = ½ × 10^9 (C/m) on the z-axis, find V_{AB} where point A is located at (2m, $\pi/2$, 0) and point B is at (4m, π, 5m).

Problem 5.10

In the field of Problem 5.9, find V_{BC} where r_B = 4m and r_c = 10m. Then, find V_{AC} (r_A = 2m) and compare with the sum of V_{AB} and V_{BC}.

Problem 5.11

Find the potential of *pt* (2, π, $\pi/2$) with respect to (4, 0, π) given the field:

$$\mathbf{E} = (-16/r^2)\ \mathbf{r}\ (\text{V/m}).$$

Problem 5.12

A line charge ρ_l = 400 pC/m lies along the x-axis and the surface of zero potential passes through the point $P_A(0, 5, 12)$m in Cartesian coordinates. Find the potential at $P_B(2, 3, -4)$m.

Problem 5.13

Find the potential at r_A = 5m with respect to r_B = 15m due to a point charge Q = 500 pC at the origin and zero reference at infinity.

Problem 5.14

A total charge of (40/3)nC is uniformly distributed around a circular ring of radius 2m.

(a) Find the potential at point on the axis 5m from the plane of the ring;

(b) Compare with the result where the full charge is at the origin in the form of a point charge.

Problem 5.15

A total charge of (40/3)nC is uniformly across a circular disk radius, r. What is the potential when r is:

(a) 1m;

(b) 2m;

(c) 4m;

(d) 8m.

Problem 5.16

Five equal point charges of $Q = 20\ \mu C$ each are located at $X = 2, 3, 4, 5$, and 6m.

Find the potential at the origin.

References

[1] Cheng, D. K., *Fundamentals of Engineering Electromagnetics*, Reading, MA: Addison-Wesley Publishing Company, Inc., 1993.

[2] Shadowitz, A., *The Electromagnetic Field*, New York: Dover Publications, Inc., 1975.

[3] Russer, P., *Electromagnetics, Microwave Circuit and Antenna Design for Communications Engineering*, Norwood, MA: Artech House, 2006.

[4] Boyce, W. E., *Calculus*, New York: John Wiley & Sons, 1988.

6

CURRENT, CURRENT DENSITY, AND CONDUCTORS

Many electrostatic concepts have been introduced in the previous five chapters. In this chapter, the subject of electric current is tackled. Along with it are Ohm's law and what occurs when charges are in motion. Conduction current density and convection current density are discussed along with the concepts of conductivity, resistance, and the continuity of current. Finally, dielectric/conductor boundary conditions in physical materials are discussed.

6.1 INTRODUCTION

Electric current is the rate of transport of electric charge past a specified point or across a specified surface. The symbol, I, is generally used for constant currents. The symbol, i, is for time-variable currents. The unit of current is the *ampere (A)* (1A = 1 C/s). In the SI unit, the ampere is the basic unit while the coulomb is the derived unit [1].

Ohm's law relates current to voltage and resistance. For simple dc circuits, $I = V/R$. However, when charges are suspended in a liquid or a gas, or where both positive and negative charge carriers are present with different characteristics, the simple form of Ohm's law is insufficient. Consequently, the current density J (A/m^2) receives more attention in electromagnetics than does current I [2].

6.2 CHARGES IN MOTION

Consider the force on a positively charged particle in a vacuum and in an electric field as shown in Figure 6.1(a). This force,

$$\mathbf{F} = +Q\mathbf{E} \quad \text{(Newtons)} \tag{6.1}$$

is unopposed and results in constant acceleration. Thus, the charge moves in the direction of **E** with a velocity **U** that increases as long as the particle is in the **E** field.

When the charge is in a liquid or gas, as shown in Figure 6.1(b), it collides repeatedly with particles in the medium, resulting in random changes in direction. But, for constant **E** and a homogeneous medium, the random velocity components cancel out, leaving a constant average velocity, known as the *drift velocity*, **U**, along the direction of **E** [1].

Conduction in metals takes place by movement of the electrons in the outermost shells of the atoms that make up the crystalline structure. According to *electron-gas theory*, these electrons reach an

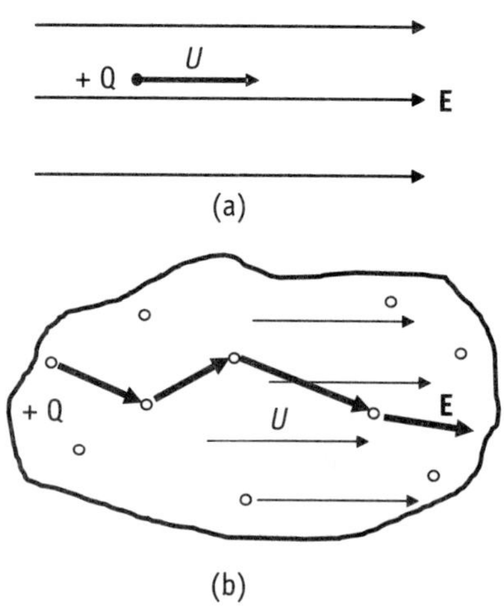

Figure 6.1 Force on positively-charged particle in: (a) vacuum, and (b) liquid or gas.

average drift velocity in much the same way as a charged particle moving through a liquid or gas. The drift velocity is directly proportional to the electric field intensity,

$$\mathbf{U} = \mu \mathbf{E} \quad (\mathrm{m}^2/\mathrm{V\text{-}s}) \tag{6.2}$$

The *mobility,* μ, has the units m^2/V-s. Each cubic meter of a conductor contains on the order of 10^{28} atoms. Good conductors have one or two electrons from each atom free to move upon application of the field [2].

The mobility, μ, varies in temperature and the crystalline structure of the solid. The particles in the solid have a vibratory motion. The vibratory motion increases with temperature, making it more difficult for the charges to move. Thus, at higher temperatures mobility, μ, is reduced, resulting in a smaller drift velocity (or current) for a given **E**. In circuit analysis, this phenomenon is accounted for by stating a *resistivity* for each material and specifying an increase in the resistivity with increasing temperature [1].

6.3 CONVECTION CURRENT DENSITY, *J*

A set of charged particles giving rise to a charge density, ρ_v, in a volume, *v*, is shown to have a velocity **U** to the right as can be seen in Figure 6.2 [3].

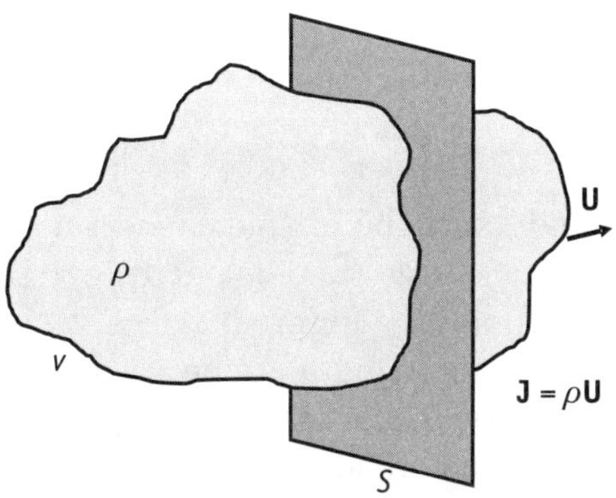

Figure 6.2 Illustration of a convection current.

The particles are assumed to maintain their relative positions within the volume. As this charge configuration passes a surface, S, it constitutes a *convection current* with density:

$$\mathbf{J} = \rho \mathbf{U} \quad (\mathrm{A/m^2}) \tag{6.3}$$

Of course, if the cross section of v varies or if the density ρ is not constant throughout v, then $\mathbf{J}$ will not be constant with time. Furthermore, $\mathbf{J}$ will be zero when the last portion of the volume crosses S. Nevertheless, the concept of a current density caused by a cloud of charged particles in motion is at times useful in the study of electromagnetic field theory [1].

6.4 CONDUCTION CURRENT DENSITY, *J*

Of more interest is the *conduction current* that occurs in the presence of an electric field within a conductor of fixed cross-section. The current density is again given by:

$$\mathbf{J} = \rho \mathbf{U} \quad (\mathrm{A/m^2}) \tag{6.4}$$

which, in view of the relation $\mathbf{U} = \mu \mathbf{E}$, can be written as

$$\mathbf{J} = \sigma \mathbf{E} \quad (\mathrm{A/m^2}) \tag{6.5}$$

where

$\sigma = \rho\mu$;

σ is the conductivity of the material in *siemens per meter (S/m)* [4].

In metallic conductors the charge carriers are electrons. These electrons (i.e., negative charge carriers), drift in a direction opposite to that of the electric field (see Figure 6.3). Hence, for electrons (negative charge carriers), both ρ and μ are negative. This in turn results in a positive conductivity, σ, just as in the case of positive charge carriers. Hence, it follows that vectors $\mathbf{J}$ and $\mathbf{E}$ have the same direction regardless of the sign of the charge carriers [2].

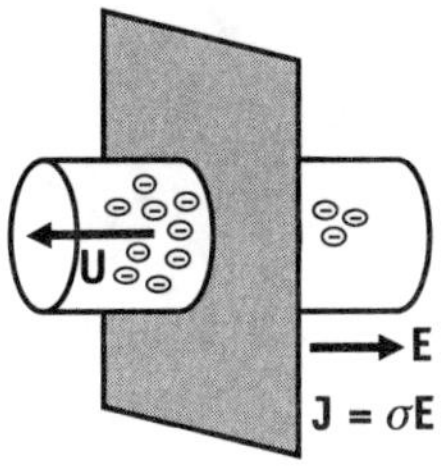

Figure 6.3 Illustration of a conduction current.

It is conventional to treat electrons as moving to the left, and positive charge carriers as moving to the right, always reporting ρ and μ as positive [3].

The relation, $\mathbf{J} = \sigma\mathbf{E}$, is often referred to as the *point form of Ohm's law*. The factor, σ, takes into account the density of the electrons being free to move (ρ) and the relative ease with which they move through the crystalline structure (μ). As might be expected, σ is also a function of temperature [1].

6.5 CONDUCTIVITY, Σ

In a liquid or gas, there are generally both positive and negative ions present, some being singly charged and others being doubly charged with possibly different masses. A conductivity expression would include all such factors. However, it is assumed that all the negative ions are alike and likewise the positive ions, then the conductivity contains two terms as shown in Figure 6.4(a) [3].

In a metallic conductor, only the valence electrons are free to move. In Figure 6.4(b) they are shown in motion going to the left. The conductivity then contains only one term, the product of the charge density of the electrons free to move, ρ_e, and their mobility, μ_e [2]. A somewhat more complex conduction occurs in semiconductor materials such as germanium and silicon. In the crystal structure, each atom has four covalent bonds with adjacent atoms.

However, at room temperature and upon influx of energy from some external source, such as light, electrons can move out of the position called for by covalent bonding to create an *electron-hole pair* available for electrical conduction. Such materials are call *intrinsic* semiconductors. Electron-hole pairs have a short lifetime,

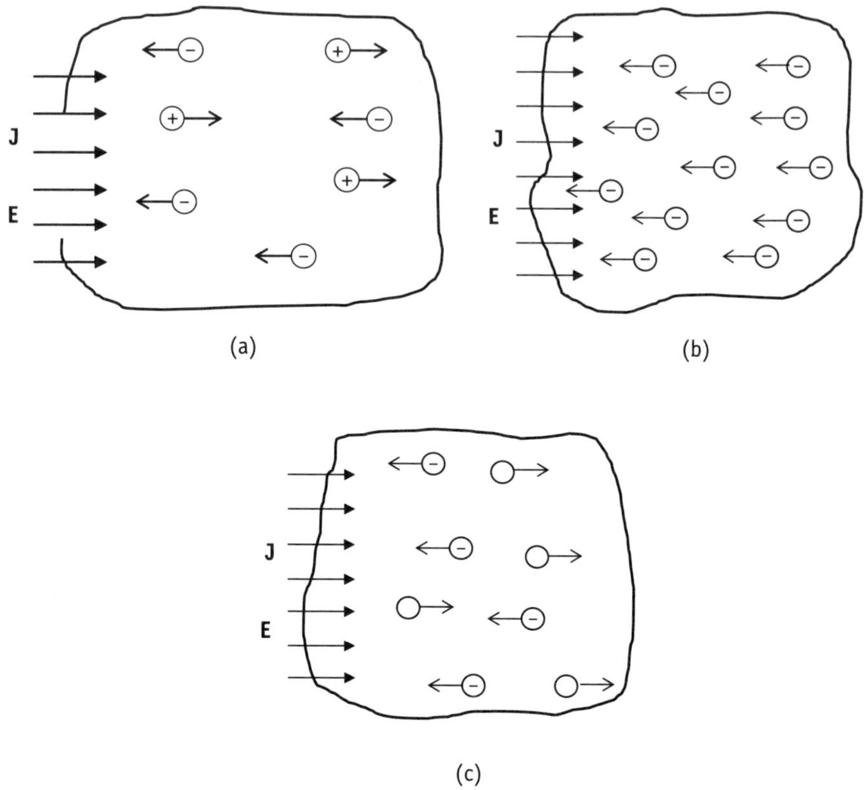

Figure 6.4 E-field and charge carrier movement in various matter types: (a) liquid or gas, (b) metallic conductor, and (c) semiconductor.

disappearing by way of *recombination*. However, other pairs are being formed constantly. Thus, at all times, some are available for conduction.

As shown in Figure 6.4(c), the conductivity, σ, consists of two terms, one for the electrons and another for the holes [4]. In practice, impurities, in the form of valence-three or valence-five elements, are added to create *p-type* and *n-type* semiconductor materials, respectively. The intrinsic behavior just described continues but is far overshadowed by the presence of extra electrons in n-type materials or holes in p-type materials. Then, in the conductivity σ, one of the densities, ρ_e or ρ_h, will far exceed the other [2].

6.6 CURRENT, *I*

The total current, I (in A), crossing a surface S is given by:

$$I = \mathbf{J} \cdot d\mathbf{S} \quad \text{(A)} \tag{6.6}$$

See Figure 6.5.

A normal vector must be chosen for the differential surface $d\mathbf{S}$. The positive result for I indicates a current crossing S in the direction of this normal vector. Of course, $\mathbf{J}$ need not be uniform over S, and S need not be a planar surface [1].

6.7 RESISTANCE, *R*

If a conductor of uniform cross-sectional area, A, and length, l, as shown in Figure 6.6, has a voltage difference, V, between its ends, then:

$$E = \frac{V}{l} \qquad \text{(V/m)} \tag{6.7}$$

and

$$J = \sigma\left(\frac{V}{l}\right) = \sigma E \quad (\text{A/m}^2) \tag{6.8}$$

where $E = V/l$.

Assuming that the current is uniformly distributed over the area A. The total current, I, is then:

$$I = JA = \frac{(\sigma A V)}{l} \quad \text{(Amperes, A)} \tag{6.9}$$

Since Ohm's law states that $V = IR$, the resistance is:

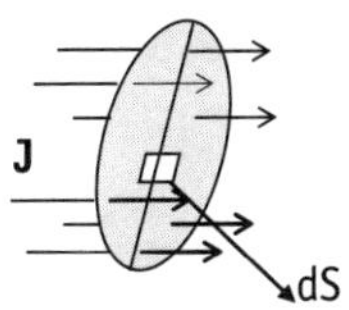

Figure 6.5 Total current crossing a surface.

$$R = \frac{l}{(\sigma A)} \quad (\Omega) \tag{6.10}$$

(Note that 1 S^{-1} = 1Ω. The siemens was formerly known as the *mho*.)

This expression for resistance is generally applied in all conductors where the cross-section remains constant over the length, *l*. However, if the current density is greater along the surface area of the conductor than in the center, then the expression is not valid. For such nonuniform current distribution, the resistance is given by:

$$R = \frac{V}{\int(\mathbf{J} \cdot d\mathbf{S})} = \frac{V}{\int(\sigma\mathbf{E} \cdot d\mathbf{S})} \quad (\Omega) \tag{6.11}$$

If **E** is known rather than the voltage difference between the two faces, the resistance is given by:

$$R = \frac{\int(\mathbf{E} \cdot d\mathbf{l})}{\int(\sigma\mathbf{E} \cdot d\mathbf{S})} \quad (\Omega) \tag{6.12}$$

The numerator gives the voltage drop across the sample, while the denominator gives the total current, *I* [2].

6.8 CURRENT SHEET DENSITY, *K*

At times current is confined to the surface of a conductor, such as inside the walls of a waveguide. For such a current sheet, it is helpful

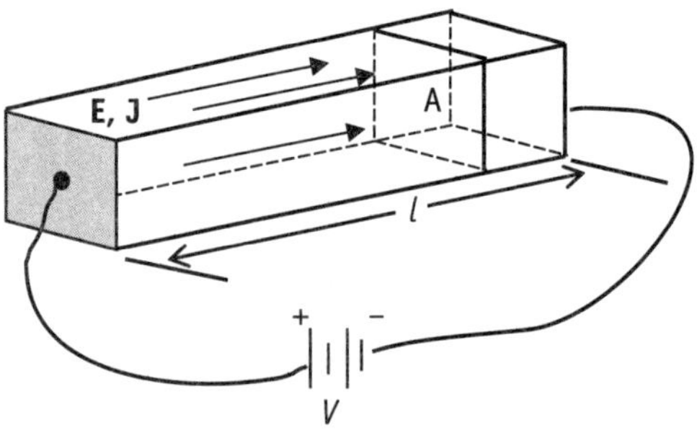

Figure 6.6 Conductor of uniform cross-sectional area with voltage potential applied between ends.

to define the density vector, **K** (in A/m), which gives the rate of charge transport per unit transverse length (note: some reference books use the notation $\mathbf{J}_S$). Figure 6.7 shows a total current of I in the form of a cylindrical sheet of radius, r, flowing in the positive z direction.

In this case,

$$\mathbf{K} = \left(\frac{I}{2\pi r}\right)\mathbf{a}_z = \left(\frac{I}{2\pi r}\right)\mathbf{z} \quad \text{(A/m)} \tag{6.13}$$

at each point of the sheet. For other sheets, **K** might vary from point to point. In general, the current flowing through a curve C within a current sheet is obtained by integrating the *normal* component of **K** along the curve (see Figure 6.8).

Thus,

$$I = \int_C K_n \, dl \quad \text{(A)} \tag{6.14}$$

6.9 CONTINUITY OF CURRENT

Current, I, crossing a general surface S has been examined where **J** at the surface was known. Now, if the surface is *closed* in order for net current to come out [5], there must be a decrease of positive charge within:

$$\oint \mathbf{J} \cdot d\mathbf{S} = I = -\frac{dQ}{dt} = -\frac{\partial}{\partial t}\int \rho \, dv \quad \text{(A)} \tag{6.15}$$

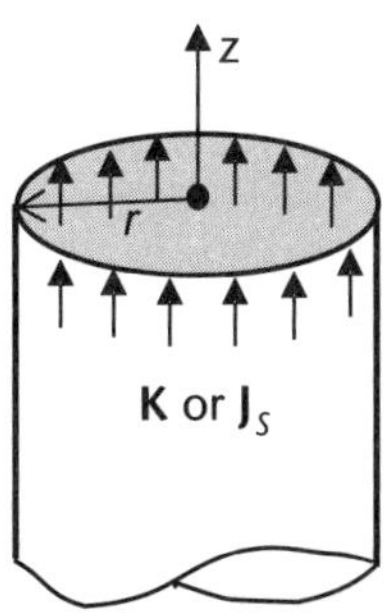

Figure 6.7 Total current, I, in the form of a cylindrical sheet of radius, r.

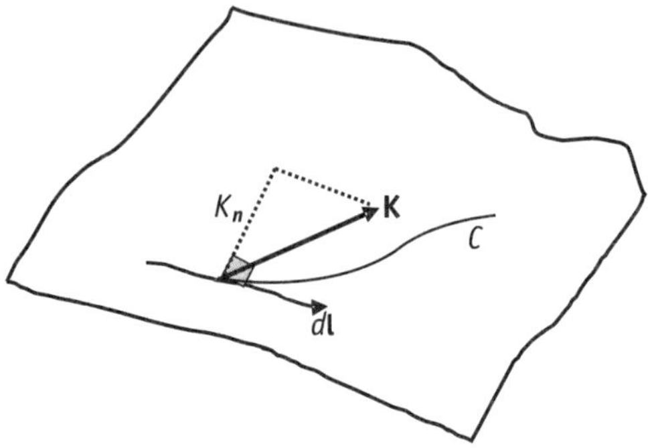

Figure 6.8 Current flowing through a curve C within a current sheet.

where the unit normal in $d\mathbf{S}$ is the outward-directed normal. Dividing by Δv,

$$\frac{\left(\oint \mathbf{J} \cdot d\mathbf{S}\right)}{\Delta v} = -\frac{\partial}{\partial t}\left[\frac{\left(\int \rho \, dv\right)}{\Delta v}\right] \tag{6.16}$$

As $\Delta v \to 0$, the left side by definition approaches $\nabla \cdot \mathbf{J}$, the divergence of the current density, while the right side approaches $-\partial\rho/\partial t$. Thus,

$$\nabla \cdot \mathbf{J} = -\frac{\partial \rho}{\partial t} \tag{6.17}$$

Equation (6.17) is the *continuity of current* equation. In it ρ stands for the *net charge* density, not just the density of mobile charge. As will be shown below, $\partial\rho/\partial t$ can be nonzero within a conductor only transiently. Then, the continuity equation, $\nabla \cdot \mathbf{J} = 0$, becomes the field equivalent of Kirchoff's current law, which states that the net current leaving a junction of several conductors is zero.

In the process of conduction, valence electrons are free to move upon the application of an electric field. So, to the extent that these electrons are in motion, static conditions no longer exist. However, these electrons should not be confused with *net charge*, for each conduction electron is balanced by a proton in the nucleus such that there is zero net charge in every Δv of the material. Suppose, however, that through a temporary imbalance a region within a solid conductor has a *net* charge density ρ_0 at time $t = 0$. Then, since

$$\mathbf{J} = \sigma\mathbf{E} = \left(\frac{\sigma}{\varepsilon}\right)\mathbf{D} \quad (\text{A/m}^2) \tag{6.18}$$

$$\nabla \cdot \left(\frac{\sigma}{\varepsilon}\right)\mathbf{D} = -\frac{\partial \rho}{\partial t} \tag{6.19}$$

The divergence operation consists of partial derivatives with respect to the spatial coordinates. If σ and ε are constants—as they would be in a homogeneous sample—then they may be removed from the partial derivatives.

$$\left(\frac{\sigma}{\varepsilon}\right)(\nabla \cdot \mathbf{D}) = -\frac{\partial \rho}{\partial t} \tag{6.20}$$

$$\left(\frac{\sigma}{\varepsilon}\right)\rho = -\frac{\partial \rho}{\partial t} \quad \text{or} \quad \left(\frac{\sigma}{\varepsilon}\right)\rho + \frac{\partial \rho}{\partial t} = 0 \tag{6.21}$$

The solution to the equation is:

$$\rho = \rho_0 e^{-t/\tau} = \rho_0 e^{-t/(\varepsilon/\sigma)} = \rho_0 e^{-(\sigma/\varepsilon)t} \tag{6.22}$$

where $\tau = \varepsilon/\sigma$ and $\varepsilon = \varepsilon_r\varepsilon_0$.

This becomes useful when knowing the relative dielectric constant of the material of interest. Also when determining charge dissipation (sometimes called charge relaxation) in liquid Jet A commercial airplane fuel that has already been filled in airplane fuel tanks. However, such fuel also has conductive (liquid) additives to increase the Jet A fuel's conductivity, σ.

It is seen that ρ, and with it

$$\frac{\partial \rho}{\partial t} = -\left(\frac{\sigma}{\varepsilon}\right)\rho \tag{6.23}$$

decays exponentially with a time constant, $\tau = \varepsilon/\sigma$, also known as the *relaxation time* for the particular material. For silver (Ag), having $\sigma = 6.17 \times 10^7$ S/m and $\varepsilon = \varepsilon_0$, the relaxation time is 1.44×10^{-19} s. Thus, if a charge density of ρ_0 could somehow be set up in the interior of a block of silver, the charges would separate due to repelling coulomb forces. After 1.44×10^{-19} s, the remaining density would be 36.8% of ρ_0. After five time-constants (or 7.20×10^{-19} s), only 0.67% of ρ_0 remains. Thus, for static fields, it may be said that the *net charge within a conductor is zero.* If any net charge is present, it must reside on the outer surface [1].

6.10 CONDUCTOR: DIELECTRIC BOUNDARY CONDITIONS

Under static conditions, all net charge will be on the outer surface of a conductor, and both **E** and **D** are therefore zero within the conductor [4]. Because the electric field is a conservative field, the closed-line integral of $\mathbf{E} \cdot d\mathbf{l}$ is zero for any path. A rectangular path with corners 1, 2, 3, 4 is shown in Figure 6.9.

Analysis:

$$\underset{1}{\int_1^2 \mathbf{E} \cdot d\mathbf{l}} + \underset{2}{\int_2^3 \mathbf{E} \cdot d\mathbf{l}} + \underset{3}{\int_3^4 \mathbf{E} \cdot d\mathbf{l}} + \underset{4}{\int_4^4 \mathbf{E} \cdot d\mathbf{l}} = 0 \quad \text{(V/m)} \tag{6.24}$$

If path length 4-to-1 and path length 2-to-3 are now allowed to approach zero while keeping the interface between them, then the second and fourth integrals in the above equation are also equal to zero. The path length from 3-to-4 is within the conductor where **E** must be zero. This leaves:

$$\int_1^2 \mathbf{E} \cdot d\mathbf{l} = \int_1^2 Et\, dl = 0$$

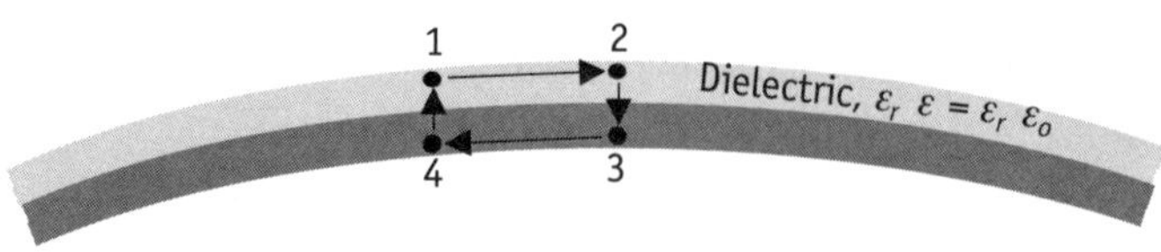

Figure 6.9 Illustration of rectangular path, adjacent dielectric, and conductor analysis: boundary conditions.

where E_t = the tangential component of **E** at the dielectric surface.

Since the interval 1-to-2 can be chosen arbitrarily,

$$E_t = D_t = 0 \tag{6.25}$$

at each point of the surface.

To discover the conditions on the normal components, a small, closed, right-circular cylinder is placed across the interface as shown in Figure 6.10.

Gauss' law applied to this surface gives:

$$\oint \mathbf{D} \cdot d\mathbf{S} = Q_{\text{enc}} \tag{6.26}$$

$$\int_{(\text{top})} \mathbf{D} \cdot d\mathbf{S} + \int_{(\text{bottom})} \mathbf{D} \cdot d\mathbf{S} + \int_{(\text{side})} \mathbf{D} \cdot d\mathbf{S} = \int_{A} \rho_S \, dS \tag{6.27}$$

The third integral is zero since, as just determined, $D_l = 0$ on either side of the interface. The second integral is also zero, since the bottom of the cylinder is within the conductor, where **D** and **E** are zero. Then,

$$\int_{(\text{top})} \mathbf{D} \cdot d\mathbf{S} + \int_{(\text{top})} D_n \cdot dS + \int_{A} \rho_S \, dS \tag{6.28}$$

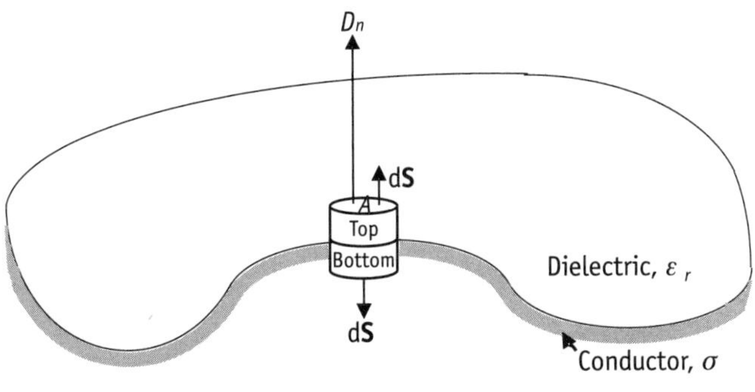

Figure 6.10 Small, closed, right-circular cylinder placed across a dielectric-conductor interface.

which can hold only if:

$$D_n = \rho_S \tag{6.29}$$

and

$$E_n = \frac{\rho_S}{\varepsilon} \tag{6.30}$$

In short, under static conditions the field just outside a conductor is zero (both tangential and normal components) unless there exists a surface charge distribution [2]. However, a surface charge does not imply a net charge in the conductor. To illustrate, consider a positive charge at the origin of spherical coordinates. Now if this point charge is enclosed by an *uncharged* conducting spherical shell of finite thickness as shown in Figure 6.11, then the field is still given by:

$$\mathbf{E} = \left(\frac{+Q}{4\pi\varepsilon r^2}\right)\mathbf{r} \quad \text{(V/m)} \tag{6.31}$$

except within the conductor itself, where **E** must be zero. The coulomb forces caused by $+Q$ attract the conduction electrons to the inner surface where they create a ρ_{s1} of negative sign. Then, the deficiency of electrons on the outer surface constitutes a positive surface charge

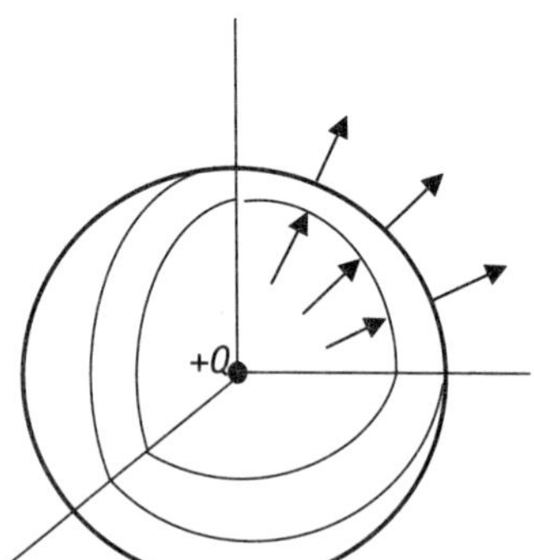

Figure 6.11 Point charge enclosed by uncharged conducting shell of finite thickness.

density, ρ_{s2}. The electric flux lines, Ψ, leaving the point charge $+Q$, terminate at the electrons on the inner surface of the conductor [4] as shown in Figure 6.12.

Then, electric flux lines Ψ originate once again on the positive charges on the outer surface of the conductor. It should be noted that the flux does not pass through the conductor, and the *net* charge on the conductor remains zero [4].

6.11 PROBLEMS AND SOLUTIONS: CURRENT, CURRENT DENSITY, AND CONDUCTORS

Problem 6.1

Find I in the circular wire (see Figure 6.13) of radius 2 mm if the current density, $\mathbf{J}$, is:

$$\mathbf{J} = 15\,(1 - e^{-1000r})\,\mathbf{z} \quad (\text{A/m}^2)$$

Problem 6.2

Find the resistance, R, between the inner and outer curved surfaces of the block shown in Figure 6.14. The material of the block is silver (Ag, $\sigma = 6.17 \times 10^7$ S/m).

Problem 6.3

An AWG #12 Cu conductor has an 80.8-mm diameter. A 50-ft length carries a current of 20A.

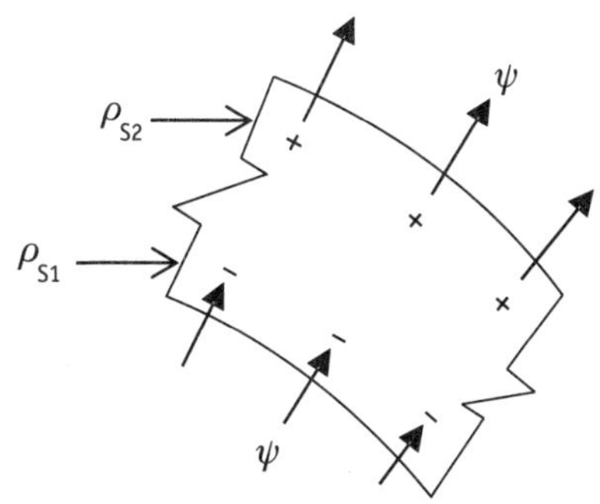

Figure 6.12 Electric flux lines, Ψ, leaving point charge $+Q$, terminating at electrons on the conductor inner surface to, once again, originate on positive charges on the conductor outside surface.

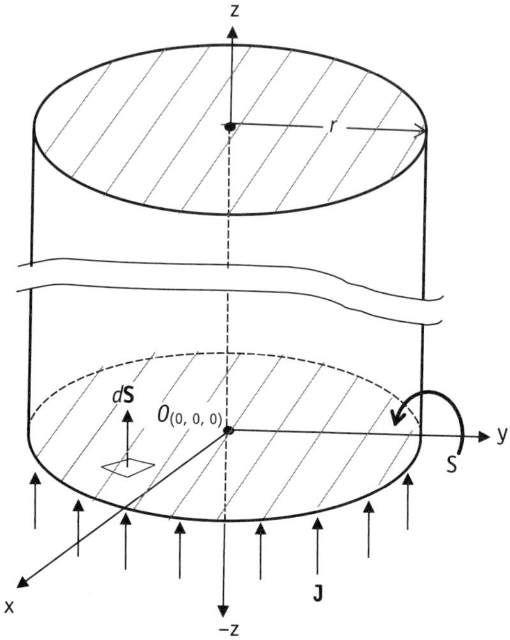

Figure 6.13 Illustration of vector analysis for Problem 6.1.

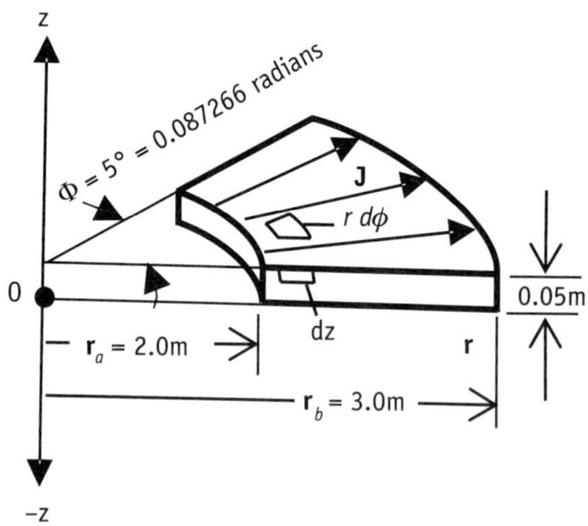

Figure 6.14 Illustration of Problem 6.2 vector analysis.

(a) What is E, the electric field intensity?

(b) What is U, the drift velocity?

(c) What is the voltage drop, V?

(d) What is the resistance, R, of the 50-ft conductor?

Problem 6.4

What current density, J, and electric field intensity, E_o, correspond to a drift velocity of 5.3×10^{-4} m/s?

Problem 6.5

A long Cu conductor has a circular cross section of diameter 3.0 mm and carries a 10A current. Each second, what percentage of the conduction electrons must leave to be replaced by others in a 100-mm length?

Problem 6.6

What current would result if all the conduction electrons in a 1-cm cube of Al passed a specified point in 2.0s? Assume one conduction electronic per atom.

Problem 6.7

What is the density of free electrons in a metal for an electron mobility, μ_M, of 0.0046 m^2/Vs and a conductivity, σ, of 29.1 MS/m?

Problem 6.8

Determine the conductivity of intrinsic Ge at room temperature.

Problem 6.9

Find the conductivity of n-type Ge at room temperature. Assume one donor atom in each batch of 10^8 atoms. The density of Ge is 5.32×10^3 kg/m^3, and the atomic weight is 72.6 kg/kmole.

Problem 6.10

A conductor of uniform cross-section and 150m length has a voltage drop of 1.3V and a current density of 4.65×10^5 A/m^2.

What is the conductivity of the material in the conductor?

Problem 6.11

A table of resistivities gives 10.4Ω CM/ft for annealed Cu.

What is the corresponding conductivity in S/m?

Problem 6.12

An AWG #20 Al wire has a resistance of 16.7 Ω/1000 ft. What conductivity for Al does this imply?

Problem 6.13

Find the current, I, crossing the portion of the $y = 0$ plane defined by:

$$-0.1\text{m} \le x \le 0.1\text{m and } -0.002\text{m} \le z \le 0.002\text{m if:}$$

$$\mathbf{J} = 10^2|x|\mathbf{y} \quad (\text{A/m}^2)$$

Problem 6.14

Find the current, I, crossing the portion of the $x = 0$ plane defined by:

$$-\pi/4 \le y \le \pi/4$$

and

$$-0.01\text{m} \le z \le 0.01\text{m}$$

if:

$$\mathbf{J} = 10^2 \cos(2y)\mathbf{x} \quad (\text{A/m}^2).$$

Problem 6.15

Find the current, I, crossing the spherical shell with $r = 0.02\text{m}$ given:

$$\mathbf{J} = 10^3 \sin\theta\, \mathbf{r} \quad (\text{A/m}^2).$$

Problem 6.16

Show that the resistance of any conductor of constant cross-sectional area, A, and length, l, is given by:

$$R = l/\sigma A,$$

assuming uniform current distribution.

Problem 6.17

A current sheet of a 4m-width lies in the $z = 0$ plane and contains a total I of 10A in a direction from the origin to $P_1(1, 3, 0)$m.

Problem 6.18

For the configuration contained in the previous problem, find the current sheet in a 30° sector of the plane.

References

[1] Shadowitz, A., *The Electromagnetic Field,* New York: Dover Publications, Inc., 1975.

[2] Russer, P., *Electromagnetics, Microwave Circuit and Antenna Design for Communications Engineering,* Norwood, MA: Artech House, 2006.

[3] Warnick, K. F., and P. Russer, *Problem Solving in Electromagnetics, Microwave Circuit and Antenna Design for Communications Engineering,* Norwood, MA: Artech House, 2006.

[4] Cheng, D. K., *Fundamentals of Engineering Electromagnetics,* Reading, MA: Addison-Wesley Publishing Company, Inc., 1993.

[5] Boyce, W. E., *Calculus,* New York: John Wiley & Sons, 1988.

7

CAPACITANCE AND DIELECTRIC MATERIALS

7.1 POLARIZATION *P* AND RELATIVE PERMITTIVITY

Dielectric materials become *polarized* in an electric field with the result that their electric flux density, D, being greater than would be under free-space conditions and under the same field intensity.

A simplified—but satisfactory—theory of polarization treats an atom of a dielectric material with two superimposed positive and negative charge regions as shown in Figure 7.1(a).

Upon applying an **E** field, the positive charge region moves in the direction of the applied field with the negative charge region moving in the direction opposite the field as illustrated in Figure 7.1(b). As a result, this displacement can be represented by an *electric dipole moment*,

$$\mathbf{p} = Q\mathbf{d} \tag{7.1}$$

as illustrated in Figure 7.1(c).

For most materials, when the applied field is removed, the charge regions will return to their original superimposed positions. Just as in the case of a spring obeying Hooke's law, the work done in the atom distortion is recovered when the system is permitted to go

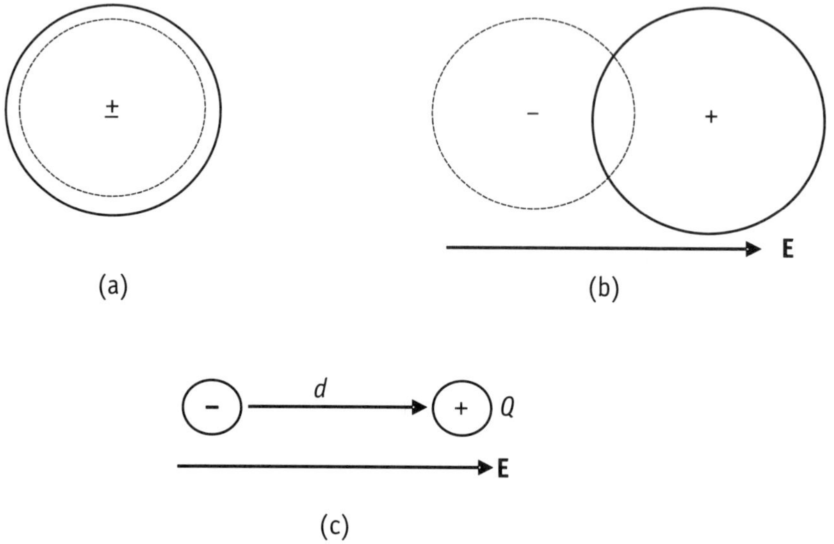

Figure 7.1 Illustration of a simplified but satisfactory theory of polarization for a dielectric: (a) atom, two superimposed, (b) positive charge region moving, and (c) resulting electric dipole positive and negative charge in the direction of the applied E-field, moment, $\mathbf{p} = Q\mathbf{d}$ regions negative-charge region moving oppositely.

back to its original material state. Energy storage taking place in this distortion is done in the same manner as with the spring.

A region Δv of a polarized dielectric will contain N-dipole moments of polarization, **p**.

Polarization **P** is defined as the dipole moment per unit volume:

$$P = \lim_{\Delta v \to 0} \left(\frac{N\mathbf{p}}{\Delta v} \right) \quad (\text{C/m}^2) \tag{7.2}$$

This definition suggests, of course, a smooth and continuous distribution of electric dipole moments throughout the volume, and this is not the case. In the macroscopic view, the polarization **P** can account for the increase in the electric flux density as shown in the following equation:

$$\mathbf{D} = \varepsilon_0 \mathbf{E} + \mathbf{P} \tag{7.3}$$

In this equation, **E** and **P** are permitted to have different directions, as they do in certain crystalline dielectrics. In an isotropic, linear material, **E** and **P** are parallel at each point. This occurrence is expressed by:

$$\mathbf{P} = \chi_e \varepsilon_0 \mathbf{E} \qquad \text{(isotropic material)} \tag{7.4}$$

Electric susceptibility, χ_e, is a dimensionless constant. Thus, we also have:

$$\mathbf{D} = \varepsilon_0 \left(1 + \chi_e\right)\mathbf{E} = \varepsilon_0 \varepsilon_r \mathbf{E} \qquad \text{(isotropic material)} \tag{7.5}$$

where $\varepsilon_r = 1 + \chi_e$ is also a pure number. Since, $\mathbf{D} = \varepsilon\mathbf{E}$,

$$\varepsilon_r = \frac{\varepsilon}{\varepsilon_0} \tag{7.6}$$

ε_r is called the *relative permittivity* [1] (compare Section 2.1).

7.2 FIXED VOLTAGE *D* AND *E*

A parallel-plate capacitor with free space between the plates and a constantly applied voltage, V, as shown in Figure 7.2, has a constant electric field intensity, **E**, between the plates.

Neglecting edge-fringing, three equations are helpful:

$$\mathbf{E} = \frac{V}{d}\mathbf{a}_n \tag{7.7}$$

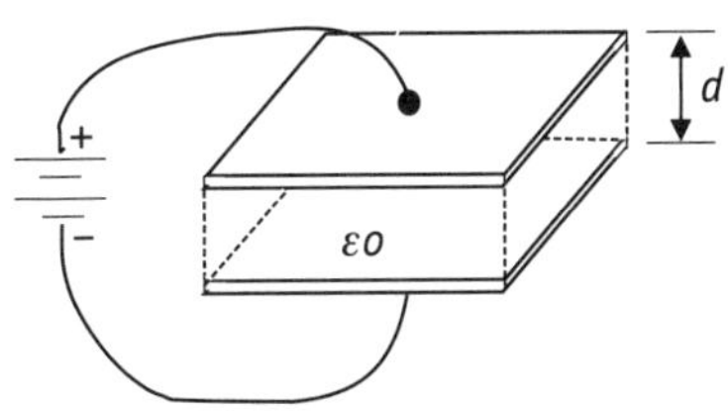

Figure 7.2 Capacitor with dielectric material between the plates and an applied voltage source.

$$\mathbf{D} = \varepsilon_0 \mathbf{E} = \varepsilon_0 \left(\frac{V}{d} \right) \mathbf{a}_n \tag{7.8}$$

$$D_n = \rho_s = \frac{Q}{A} \tag{7.9}$$

So, when a dielectric with relative permittivity, ε_r, fills the space between the plates,

$$\mathbf{D} = \varepsilon_0 \mathbf{E} + \mathbf{P} = \varepsilon_0 \mathbf{E} + \varepsilon_0 \chi_e \mathbf{E} \tag{7.10}$$

the equations are:

$$\mathbf{E} = \frac{V}{d} \mathbf{a}_n \quad \text{(as in free space)} \tag{7.11}$$

$$\mathbf{D} = \varepsilon_0 \varepsilon_r \mathbf{E} \tag{7.12}$$

Since $D_n = \rho_s = Q/A$, the charge and charge density increase by the factor, ε_r, over their free-space values. This charge increase is supplied by the source voltage, V [2].

7.3 FIXED CHARGE *D* AND *E*

The parallel-plate capacitor in Figure 7.3 has a charge $+Q$ on the upper plate and $-Q$ on the lower plate.

The charges could have been the result of a connected voltage source, V, which was subsequently removed. With free space between the plates and fringing neglected,

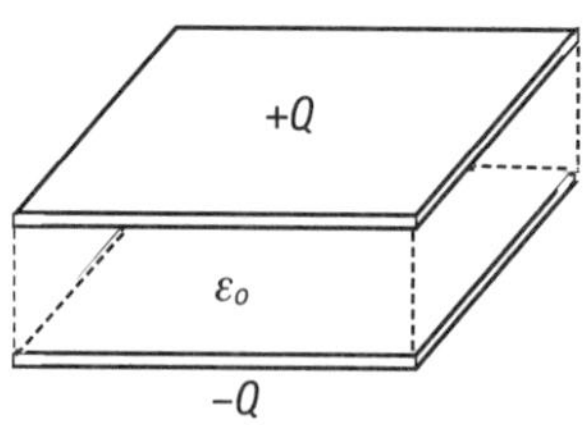

Figure 7.3 Capacitor with dielectric material between the plates and positive charge on the top plate and negative charge at the bottom plate.

$$D_n = \rho_s = \frac{Q}{A} \tag{7.13}$$

$$\mathbf{E} = \frac{\mathbf{D}}{\varepsilon_0} = \left(\frac{\rho_s}{\varepsilon_0}\right)\mathbf{a}_n \tag{7.14}$$

In this arrangement, there is no way for the charge to increase or decrease since, ideally (i.e., ideal dielectric material), there is no conducting path to the plates. When a dielectric material is assumed to fill the space between the plates, the equations are:

$$D_n = \rho_s = \frac{Q}{A} \tag{7.15}$$

$$\mathbf{E} = \frac{\mathbf{D}}{\varepsilon_0 \varepsilon_r} \tag{7.16}$$

With a constant Q and ρ_s, **D** must be the same as under free-space conditions whereas the magnitude of **E** decreases by the factor, $1/e_r$. The decrease in **E** ε_0 is made up for by polarization, **P**, in the relation, $\mathbf{D} = \varepsilon_0\mathbf{E} + \mathbf{P}$. More generally, a homogeneous dielectric of relative permittivity, ε_r, the coulomb force between the charges is reduced to $1/e_r$ of its free-space value [1]:

$$\mathbf{F} = \left(\frac{Q_1 Q_2}{4\pi\varepsilon d^2}\right)\mathbf{a} = \left(\frac{1}{\varepsilon_r}\right)\left(\frac{Q_1 Q_2}{4\pi\varepsilon_0 d^2}\right)\mathbf{a} \tag{7.17}$$

7.4 BOUNDARY CONDITIONS AT THE INTERFACE OF TWO DIELECTRICS

If the conductor in Figures 6.9 and 6.10 is replaced by a second, and different dielectric, then the same argument, as was made in Section 6.10, can be made. In doing so, the following two boundary conditions are established:

(a) The tangential components of **E** are continuous across a dielectric interface. In symbols, this is:

$$E_{t1} = E_{t2} \tag{7.18}$$

and

$$\frac{D_{t1}}{\varepsilon_{r1}} = \frac{D_{t2}}{\varepsilon_{r2}} \tag{7.19}$$

(b) The normal component of **D** has a discontinuity of magnitude $|\rho_s|$ across a dielectric interface. If the normal unit vector is chosen to point into dielectric 2—from dielectric 1—then the following conditions can be written as:

$$D_{n1} - D_{n2} = -\rho_s \tag{7.20}$$

and

$$\varepsilon_{r1}E_{n1} - \varepsilon_{r2}E_{n2} = -\frac{\rho_s}{\varepsilon_0} \tag{7.21}$$

Generally speaking, the interface will have no free charges in it ($\rho_s = 0$) such that:

$$D_{t1} = D_{t2} \tag{7.22}$$

and

$$\varepsilon_{r1}E_{n1} = \varepsilon_{r2}E_{n2} \tag{7.23}$$

7.5 CAPACITANCE

Any two conducting bodies, separated by a dielectric material or free space (no dielectric material), have a capacitance that exists between them. A voltage difference applied between the two conductive bodies results in a $+Q$ charge on one conductor and $-Q$ on the other. The ratio of the absolute value of the charge to the absolute value of the voltage potential difference is defined as the capacitance of the system:

$$C = \frac{Q}{V} \text{ (Farads, F)}, \tag{7.24}$$

where 1 Farad (F) = 1 C/V.

The capacitance depends only on the geometry of the system and the properties of the dielectric(s) involved. In Figure 7.4, $+Q$ charge is placed on the upper conductor (#1), and $-Q$ charge is placed on the lower conductor (#2). The result is that a flux (E-) field is created. Both **D** and **E** fields are established together. Doubling both $+Q$ and $-Q$ simply doubles the magnitude of **D** and **E**, and doubles also the voltage potential difference between the conductive plates. Hence, the ratio Q/V would remain fixed [2].

7.6 MULTIPLE-DIELECTRIC CAPACITORS

When two dielectrics are present with the interface parallel to **E** and **D** as in Figure 7.5(a), the capacitance can be found by treating the arrangement as two capacitors in parallel (see Figure 7.5(a, b)).

Of course, the result may be extended to any number of side-by-side dielectrics. Hence, the equivalent capacitance is the sum of the two individual capacitances.

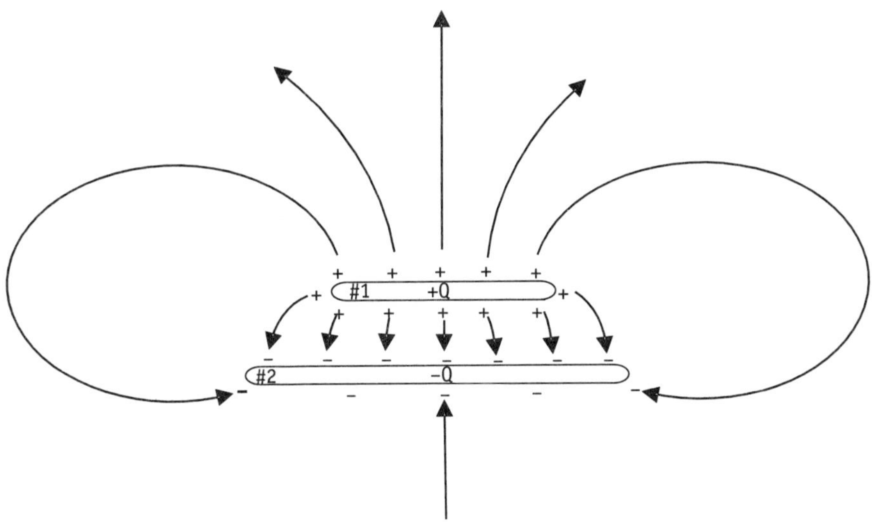

Figure 7.4 Capacitor: two plates of unequal width separated by a distance, showing paths and the direction of the E-fields when voltage is applied.

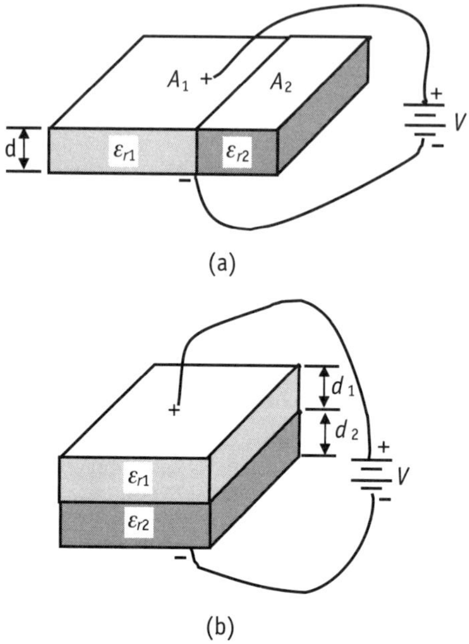

Figure 7.5 Capacitor example, two adjacent dielectric materials arranged in two different ways: (a) with the interface parallel and (b) with the interface normal.

When the dielectric interface is normal to **D** and E, as in Figure 7.5(b), the capacitance can be found by treating the arrangement as two capacitors in series, hence:

$$\frac{1}{C_{\text{equiv}}} = \frac{1}{C_1} + \frac{1}{C_2} \tag{7.25}$$

The result can be extended to any number of stacked dielectrics:

The reciprocal of the equivalent capacitance is the sum of the reciprocals of the individual capacitances [2].

7.7 ENERGY STORED IN A CAPACITOR

The energy stored in a capacitor is given by the expression:

$$W_E = \frac{1}{2}\int \mathbf{D} \cdot \mathbf{E}\, dv \tag{7.26}$$

Where the integration may be taken over is the space between the conductors (with the fringing neglected). If this space is occupied by a dielectric of relative permittivity, ε_r, then

$$\mathbf{D} = \varepsilon_0 \mathbf{E} + \mathbf{P} = \varepsilon_0 \varepsilon_r \mathbf{E} \tag{7.27}$$

and so

$$W_E = \frac{1}{2}\int \left(\varepsilon_0 E^2 + \mathbf{P} \cdot \mathbf{E}\right)\, dv = \frac{1}{2}\int \varepsilon_0 \varepsilon_r E^2 \, dv \tag{7.28}$$

The two above expressions show how the presence of a dielectric results in an increase in stored energy over the free-space value ($\mathbf{P} = 0$, $\varepsilon_r = 1$) either through the term $\mathbf{P} \cdot \mathbf{E}$ or through the factor $\varepsilon_r > 1$ [4]. Given in terms of the capacitance,

$$W_E = \frac{1}{2} C V^2 \tag{7.29}$$

and the effect of the dielectric is reflected in C which is directly proportional to ε_r [1].

7.8 PROBLEMS AND SOLUTIONS: CAPACITANCE AND DIELECTRIC MATERIALS

Problem 7.1

Find the capacitance of a coaxial capacitor of length, h, where the outer surface of the inner conductor has a radius, a, and the inner surface of the outer conductor has a radius, b. See Figure 7.6.

Problem 7.2

Find the capacitance of the parallel plates (see Figure 7.7). Neglect fringing.

Problem 7.3

Find the polarization **P** in a dielectric material with $e_r = 2.8$ if:

$$\mathbf{D} = 3.0 \times 10^{-7}\, \mathbf{a} \quad \mathrm{C/m^2}$$

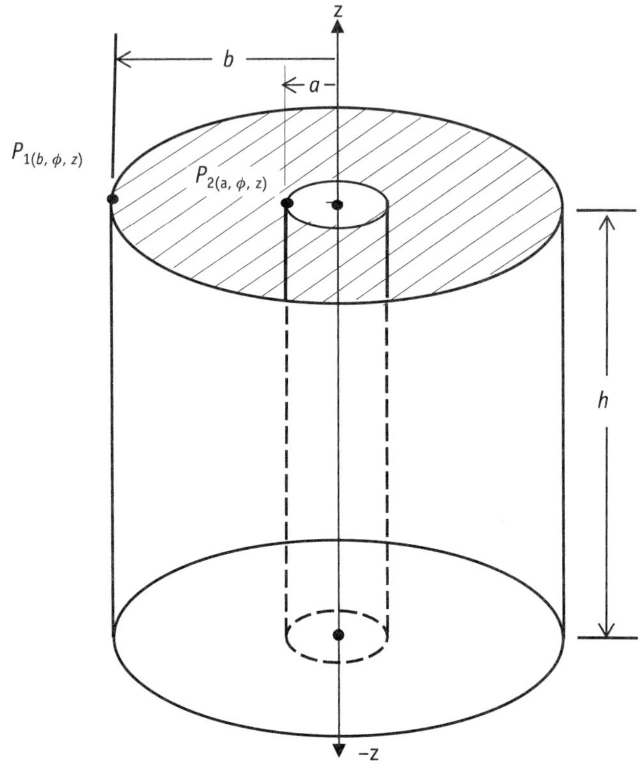

Figure 7.6 Illustration of vector analysis for Problem 7.1.

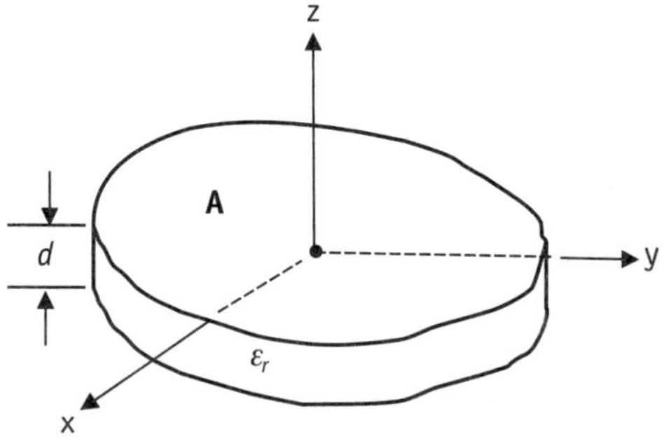

Figure 7.7 Parallel plates with dielectric material in between them.

Problem 7.4

Determine the value of **E** in a material with an electric susceptibility of 3.5 and:

$$\mathbf{P} = 2.30 \times 10^{-7}\ \mathbf{a} \quad \text{C/m}^2.$$

Problem 7.5

Two point charges in a dielectric medium ($e_r = 5.2$) interact with a force of 8.6×10^{-3} N. What force could be expected if the charges were in free space?

Problem 7.6

Region 1, shown in Figure 7.8, is defined by $x < 0$ and is in free space. Region 2 in the figure, defined by $x > 0$, is a dielectric material with an $\varepsilon_r = 2.4$.

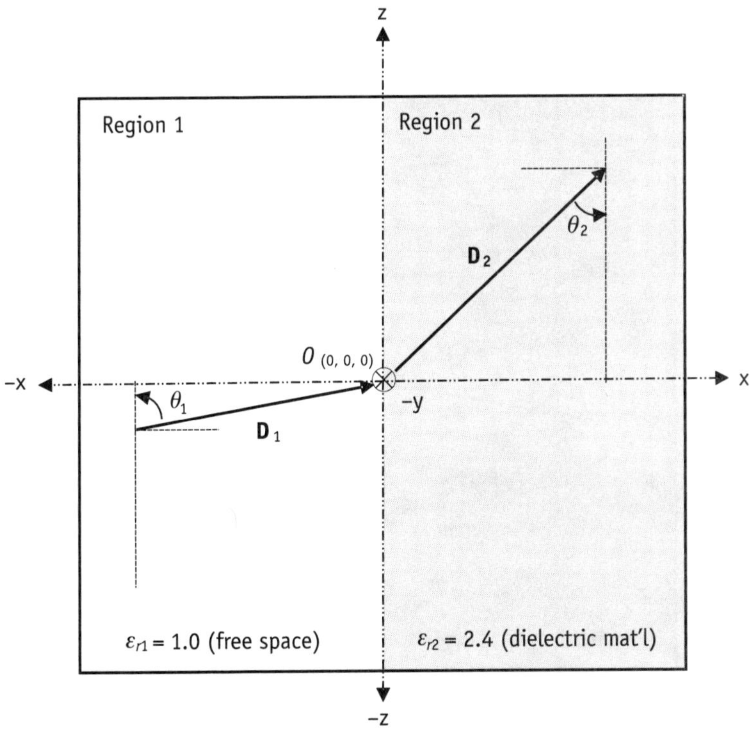

Figure 7.8 Free space/dielectric material interface, region 1 and region 2.

Given $\mathbf{D}_1 = 3\mathbf{x} - 4\mathbf{y} + 6\mathbf{z}$, find $\mathbf{E}_2$, find angles θ_1 and θ_2.

Problem 7.7

In the free-space region (#1) where $x < 0$,

$$\mathbf{E}_1 = 3\mathbf{x} + 5\mathbf{y} - 3\mathbf{z} \quad \text{V/m.}$$

In the region (#2) where $x > 0$ is a dielectric material of $\varepsilon_r = 3.6$. Find θ_2 that the E-field makes in the dielectric material with the $x = 0$ plane (see Figure 7.9).

Problem 7.8

A dielectric-free space interface (i.e., a plane) is characterized by the equation:

$$3x + 2y + z = 12\text{m}.$$

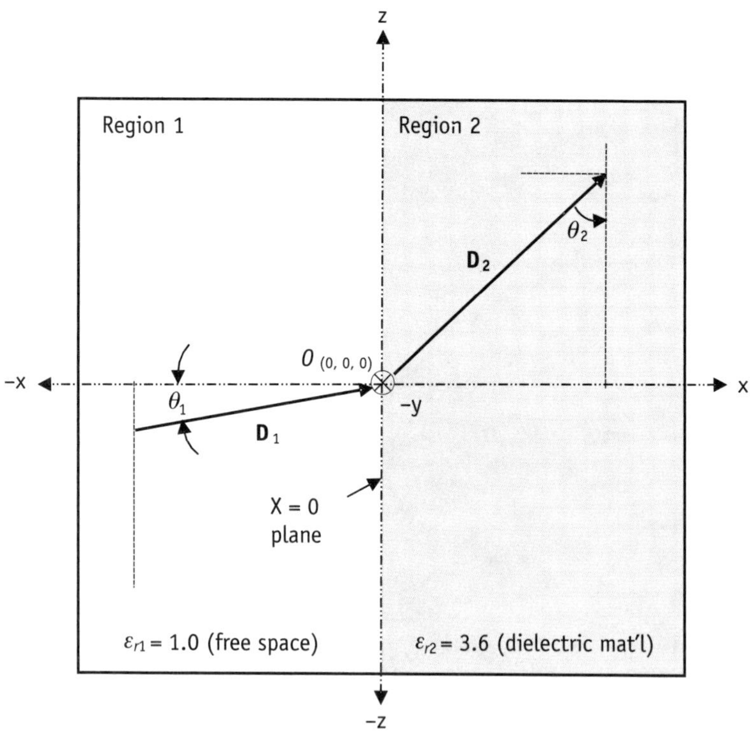

Figure 7.9 Free-space/dielectric material interface for Problem 7.7.

The origin side of the interface has $\varepsilon_r = 3.0$ and an E-field of:

$$\mathbf{E}_1 = 2\mathbf{x} + 5\mathbf{z} \quad \text{V/m}$$

Find E_2. See Figure 7.10.

Problem 7.9

Figure 7.11 shows a planar dielectric slab with free space on either side. Show that $\mathbf{E}_3 = \mathbf{E}_1$, assuming a constant field $\mathbf{E}_2$ with the slab.

Problem 7.10

Show that the capacitor of Figure 7.12 has capacitance:

$$C_{eq} = \varepsilon_0\varepsilon_{r1}\,(A_1/d) + \varepsilon_0\varepsilon_{r2}\,(A_2/d) = C_1 + C_2$$

Problem 7.11

Show that the capacitor of Figure 7.13 has reciprocal capacitance,

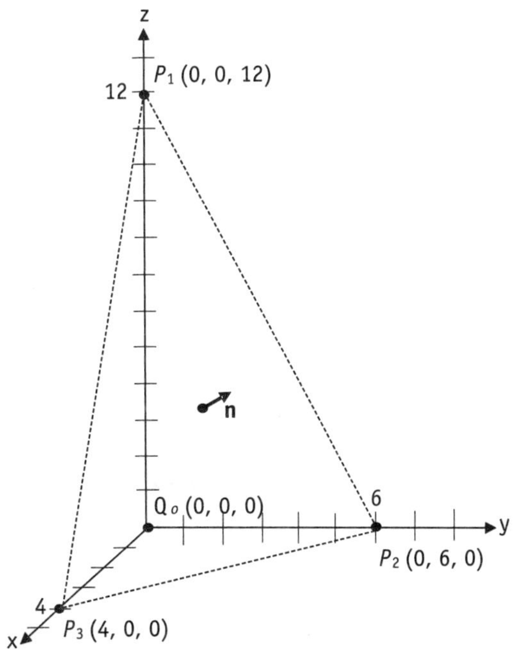

Figure 7.10 Dielectric-free space interface for Problem 7.8.

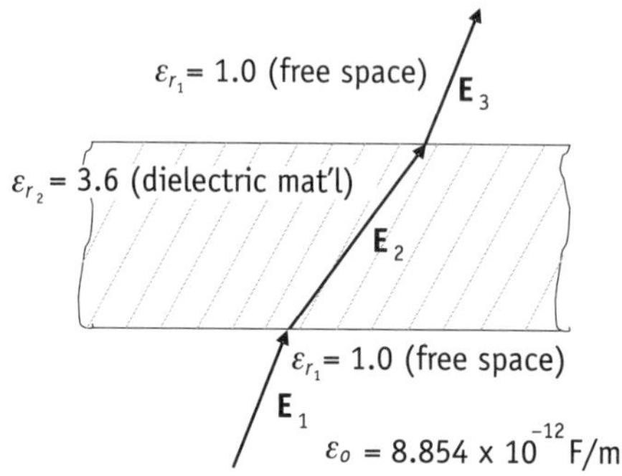

Figure 7.11 Free-space/dielectric material interface for Problem 7.9.

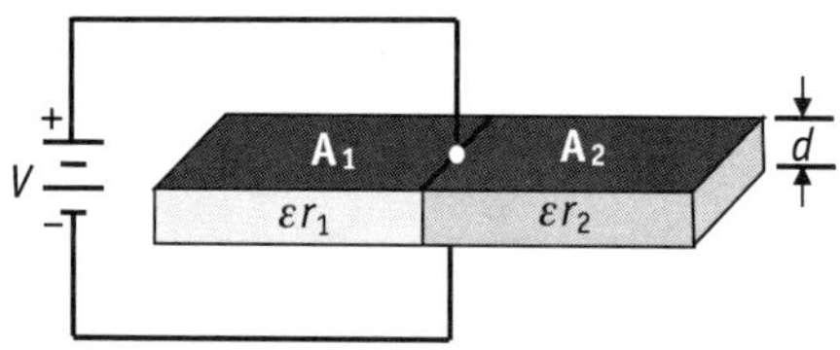

Figure 7.12 Free-space/dielectric material interface for Problem 7.10.

$$1/C_{eq} = 1/[\varepsilon_0\varepsilon_{r1}(A_1/d)] + 1/[\varepsilon_0\, \varepsilon_{r2}\, (A_2/d)] = 1/C_1 + 1/C_2$$

Problem 7.12

Find the capacitance of a coaxial capacitor of length, L, where the inner conductor has a radius, a, and the outer conductor has a radius, b (see Figure 7.14).

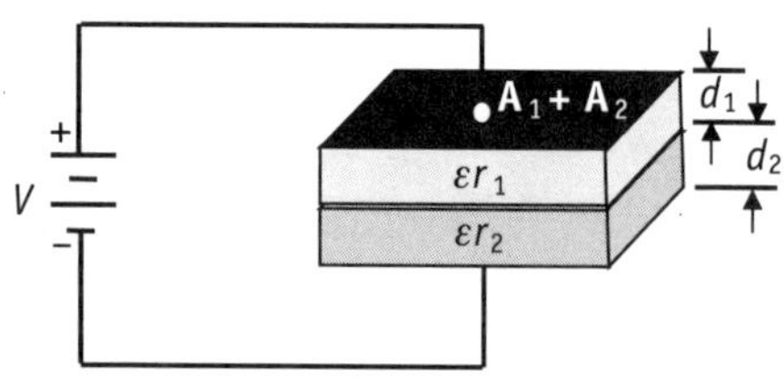

Figure 7.13 Free-space/dielectric material interface for Problem 7.11.

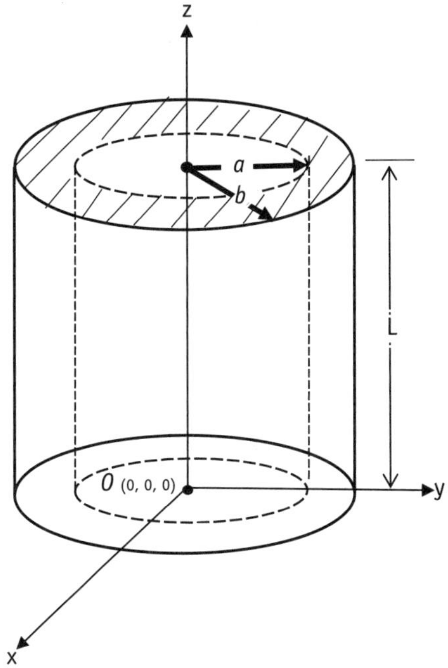

Figure 7.14 Free-space/dielectric material interface between cylindrical conductors.

Problem 7.13

In the capacitor of Figure 7.15, the region between the plates is filled with a dielectric of $\varepsilon_r = 4.5$.

Find the capacitance (see Figure 7.15).

Problem 7.14

Find the capacitance of an isolated spherical shell of radius, a (see Figure 7.16).

Problem 7.15

Find the capacitance between the two spherical shells of radius, a, separated by a distance, $d >> a$ (see Figure 7.17).

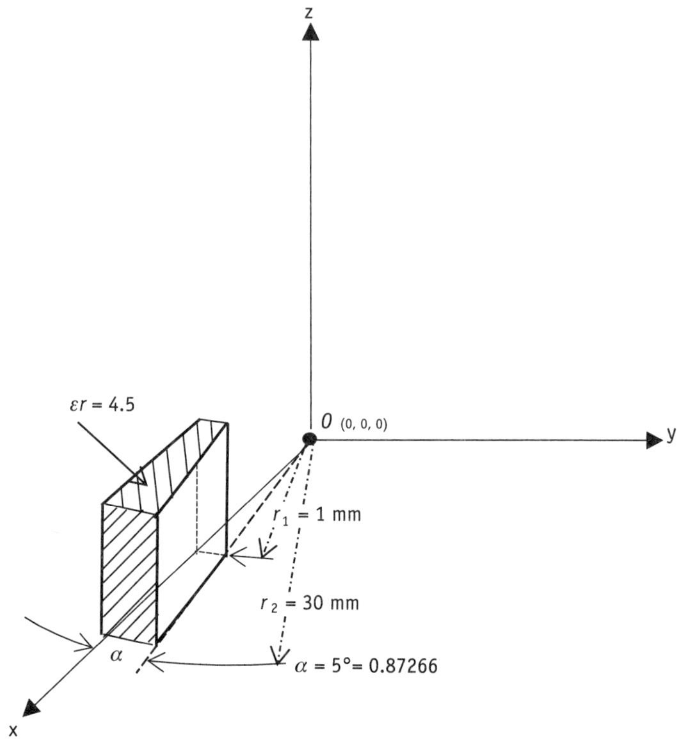

Figure 7.15 Free-space/dielectric material interface for Problem 7.13.

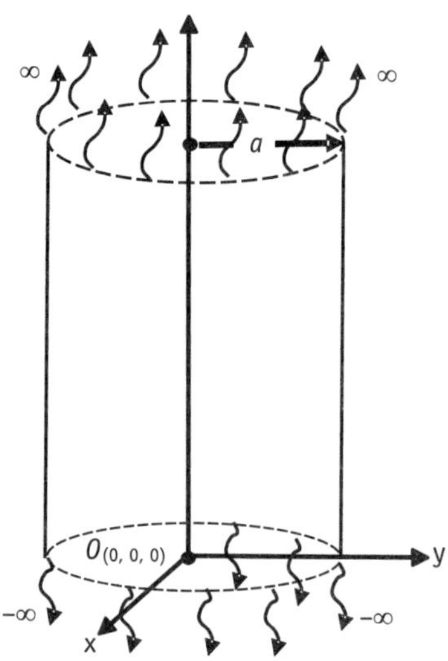

Figure 7.16 Free-space/dielectric material interface for Problem 7.14.

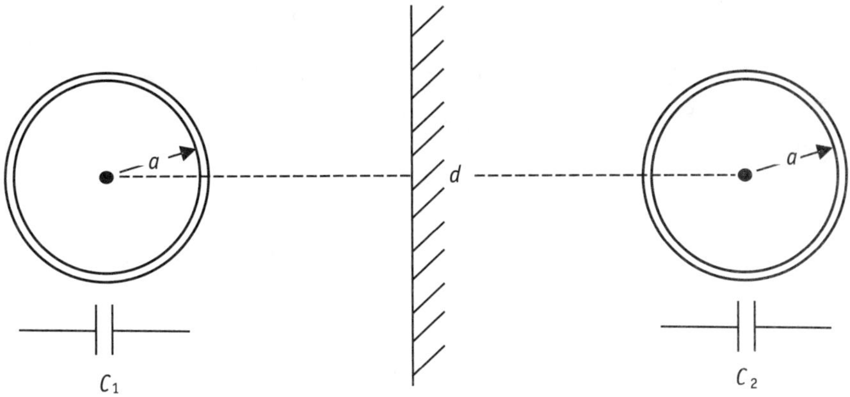

Figure 7.17 Free-space/dielectric material interface for Problem 7.15.

Problem 7.16

Find the capacitance of a parallel-plate capacitor containing two dielectrics, $\varepsilon_{r1} = 1.5$ and $\varepsilon_{r2} = 3.5$, each comprising one-half the total volume as shown in Figure 7.18.

Problem 7.17

Repeat the previous problem except for the parallel plates and two dielectrics, $\varepsilon_{r1} = 1.5$ and $\varepsilon_{r2} = 3.5$, as shown in Figure 7.19.

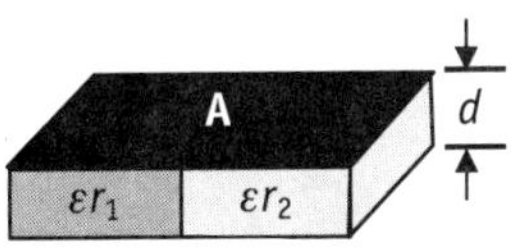

Figure 7.18 Free-space/dielectric material interface for Problem 7.16.

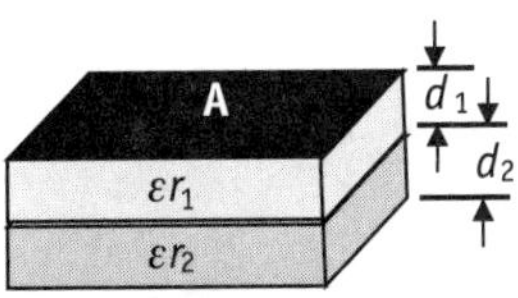

Figure 7.19 Free-space/dielectric material interface for Problem 7.17.

Problem 7.18

Each dielectric in the cylindrical capacitor (and there are two of them, see Figure 7.20) occupies one half of the volume. Find the capacitance.

$$C = 2\varepsilon_{r\text{-ave}}\,[(\pi\varepsilon_0 L)/\ln(b/a)] = [(2\pi\varepsilon_0 L)/\ln(b/a)]\,\varepsilon_{r\text{-ave}}$$

Problem 7.19

Find the voltage across each dielectric in the capacitor shown in Figure 7.21 when the applied voltage is 200V.

Problem 7.20

Find the voltage drop across each dielectric in Figure 7.22 where $\varepsilon_{r1} = 2.0$ and $\varepsilon_{r2} = 5.0$. The inner conductor is $r_1 = 2$ cm and the outer

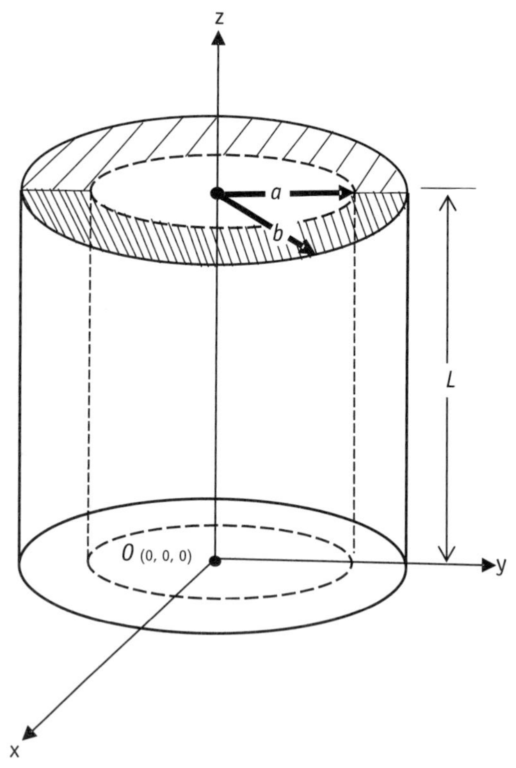

Figure 7.20 Free-space/dielectric material interface for Problem 7.18.

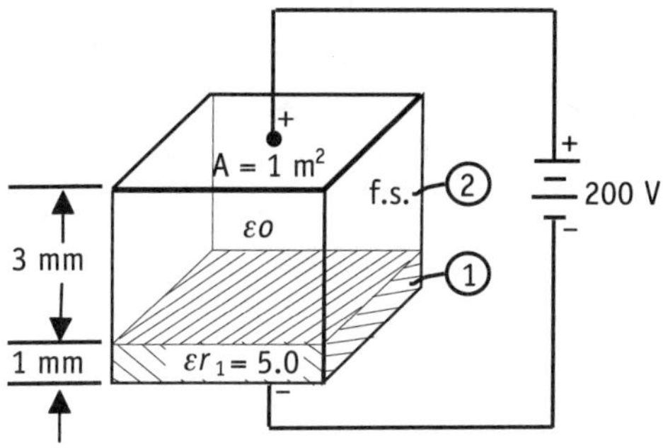

Figure 7.21 Free-space/dielectric material interface for Problem 7.19.

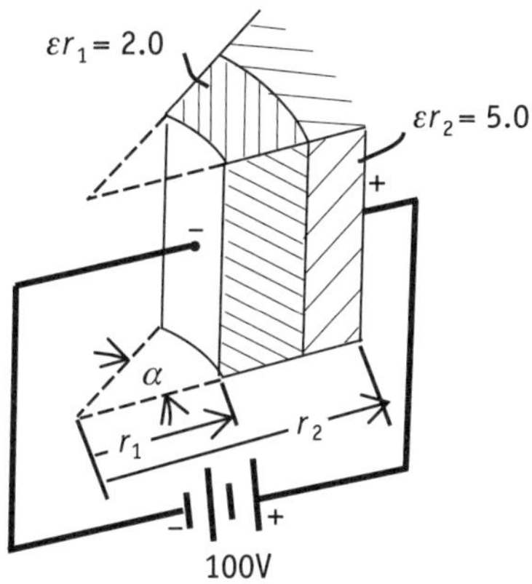

Figure 7.22 Free-space/dielectric material interface for Problem 7.20.

conductor is $r_2 = 2.5$ cm with dielectric interface that is located halfway between them.

References

[1] Russer, P., *Electromagnetics, Microwave Circuit and Antenna Design for Communications Engineering*, Norwood, MA: Artech House, 2006.

[2] Shadowitz, A., *The Electromagnetic Field*, New York: Dover Publications, Inc., 1975.

[3] Warnick, K. F., and P. Russer, *Problem Solving in Electromagnetics, Microwave Circuit and Antenna Design for Communications Engineering*, Norwood, MA: Artech House, 2006.

[4] Cheng, D. K., *Fundamentals of Engineering Electromagnetics*, 4th Edition, McGraw-Hill Education, 2019.

8

LAPLACE'S EQUATION

8.1 INTRODUCTION

Electric field intensity, **E**, was discussed in Chapter 2 and determined by summation or integration of point charges, line charges, and other charge configurations. In Chapter 3, Gauss' law was used to obtain **D**, which then gives **E**. While these two approaches are of value to an understanding of electromagnetic (EM) theory, they both tend to be impractical because charge distributions are not usually known. On the other hand, the method of Chapter 5, where **E** was found to be the negative of the gradient of V, requires that the potential function throughout the region be known. However, unfortunately, the potential function is generally not known.

Instead, conducting materials in the form of planes, curved surfaces, or lines are usually specified, and the voltage on one is also known with respect to some reference voltage that is located on one of the other conductors. Laplace's equation then provides a method whereby the potential function, V, can be obtained subject to the boundary conditions on the conductors [1].

8.2 POISSON'S EQUATION AND LAPLACE'S EQUATION

In Section 4.3, one of Maxwell's equations, $\nabla \mathbf{D} = \rho$, was developed. By substituting $\varepsilon \mathbf{E} = \mathbf{D}$ and $-\nabla V = \mathbf{E}$, we have:

$$\nabla \cdot \varepsilon(-\nabla V) = \rho \tag{8.1}$$

If the medium is homogeneous throughout the region of interest, ε may then be removed from the partial derivatives involved in the divergence. Hence, this gives:

$$\nabla \cdot \nabla V = -\frac{\rho}{\varepsilon} \quad \text{or} \quad \nabla^2 V = -\frac{\rho}{\varepsilon} \tag{8.2}$$

which is *Poisson's equation*. When the region of interest contains charges in a known distribution, ρ, Poisson's equation can be used to determine the potential function. However, the region very often is charge-free and has uniform permittivity. Poisson's equation then becomes *Laplace's equation* [2]:

$$\nabla^2 V = 0 \tag{8.3}$$

8.3 EXPLICIT FORMS OF LAPLACE'S EQUATION

Since the left side of Laplace's equation is the *divergence of the gradient of* V, these operations can be used to arrive at the form of the equation in a particular coordinate system. Mathematical displays of the divergence of the gradient of V are given, for each of the three coordinate systems (i.e., Cartesian, cylindrical, and spherical systems), as follows:

Cartesian coordinates:

$$\nabla V = \frac{\partial V}{\partial x}\mathbf{a}_x + \frac{\partial V}{\partial y}\mathbf{a}_y + \frac{\partial V}{\partial z}\mathbf{a}_z \tag{8.4}$$

For general vector field, $\mathbf{A}$,

$$\nabla \cdot \mathbf{A} = \frac{\partial A_x}{\partial x} + \frac{\partial A_y}{\partial y} + \frac{\partial A_z}{\partial z} \tag{8.5}$$

Hence, Laplace's equation in the Cartesian coordinate system, is [1]:

$$\nabla^2 V = \frac{\partial^2 V}{\partial x^2} + \frac{\partial^2 V}{\partial y^2} + \frac{\partial^2 V}{\partial z^2} = 0 \tag{8.6}$$

Cylindrical coordinates:

$$\nabla V = \frac{\partial V}{\partial r}\mathbf{a}_r + \left(\frac{1}{r}\right)\frac{\partial V}{\partial \varphi}\mathbf{a}_\varphi + \frac{\partial V}{\partial z}\mathbf{a}_z \tag{8.7}$$

For general vector field **A**,

$$\nabla \cdot \mathbf{A} = \left(\frac{1}{r}\right)\frac{\partial}{\partial x}(rA_r) + \left(\frac{1}{r}\right)\frac{\partial A_\varphi}{\partial \varphi} + \frac{\partial A_z}{\partial z} \tag{8.8}$$

Hence, Laplace's equation in the cylindrical coordinate system is [1]:

$$\nabla^2 V = \left(\frac{1}{r}\right)\frac{\partial}{\partial r}\left(\frac{r\,\partial V}{\partial \varphi}\right) + \left(\frac{1}{r^2}\right)\left(\frac{\partial^2 V}{\partial \varphi^2}\right) + \frac{\partial^2 V}{\partial z^2} = 0 \tag{8.9}$$

Spherical coordinates:

$$\nabla V = \frac{\partial V}{\partial r}\mathbf{a}_r + \left(\frac{1}{r}\right)\frac{\partial V}{\partial \varphi}\mathbf{a}_\varphi + \left(\frac{1}{r\sin\theta}\right)\frac{\partial V}{\partial \theta}\mathbf{a}_\theta \tag{8.10}$$

For a general vector field **A**,

$$\nabla \cdot \mathbf{A} = \left(\frac{2}{r^2}\right)\frac{\partial}{\partial r}(r^2 A_r) + \left(\frac{1}{r\sin\theta}\right)\frac{\partial}{\partial \phi}(A_\phi \sin\theta) + \left(\frac{1}{r\sin\theta}\right)\frac{\partial A_\phi}{\partial \phi} \tag{8.11}$$

Hence, Laplace's equation in the spherical coordinate system is [1]:

$$\nabla^2 V = \left(\frac{1}{r^2}\right)\frac{\partial}{\partial r}\left(\frac{r^2\,\partial V}{\partial r}\right) + \left(\frac{1}{r^2\sin\theta}\right)\frac{\partial}{\partial \theta}\left[\sin\theta\left(\frac{\partial V}{\partial \theta}\right)\right] + \left(\frac{1}{r^2\sin^2\theta}\right)\frac{\partial^2 V}{\partial \phi^2} = 0 \tag{8.12}$$

8.4 UNIQUENESS THEOREM

Any solution to either Laplace's or Poisson's equation that also satisfies the boundary conditions must be the only existing solution. Thus, that solution is said to be *unique*. Sometimes, there is confusion on this point. However, usually, that is due mainly to incomplete boundaries.

Consider the conducting plane at $z = 0$, shown in Figure 8.1, having a voltage of 100V.

It is clear that both:

$$V_1 = 5z + 100 \tag{8.13}$$

and

$$V_2 = 100 \tag{8.14}$$

satisfy Laplace's equation and the requirement that $V = 100$ when $z = 0$. The answer is that a single conducting surface with a specified voltage and no reference voltage given does not form the complete boundary of a properly defined region [3]. Even two finite parallel conducting planes do not form a complete boundary since the fringing of the field around the edges cannot be determined.

However, when parallel planes are specified and when it is stated to neglect fringing, then the region between the planes has boundaries that are properly defined.

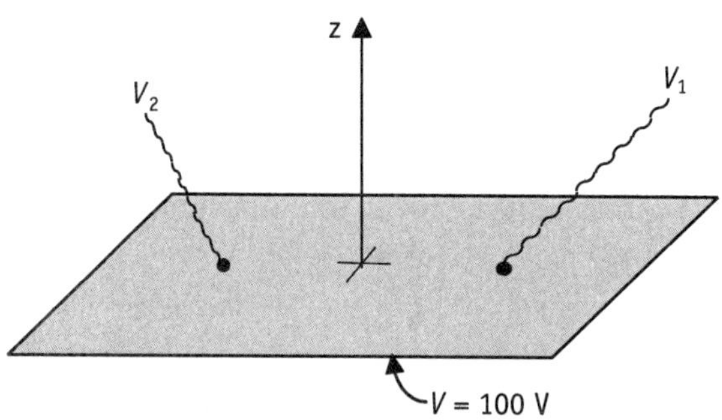

Figure 8.1 Conducting plane at $z = 0$.

8.5 MEAN VALUE AND MAXIMUM VALUE THEOREMS

Two important properties of the potential in a charge-free region can be obtained from Laplace's equation:

1. At the center of an included circle or sphere, the potential difference, V, is equal to the average of the values it assumes on the circle or sphere;
2. The potential difference, V, cannot have a maximum (or a minimum) within the region.

It then follows from the second above-stated property that any maximum of V must occur on the other boundary region. Since V obeys Laplace's equation:

$$\frac{\partial^2 V}{\partial x^2} + \frac{\partial^2 V}{\partial y^2} + \frac{\partial^2 V}{\partial z^2} = 0 \tag{8.15}$$

so do $\partial V/\partial x$, $\partial V/\partial y$, and $\partial V/\partial z$. Thus, *the components of the electric field intensity take their maximum values on the boundary* [4].

8.6 CARTESIAN SOLUTION IN ONE VARIABLE

Consider the parallel conductors of Figure 8.2.

In Figure 8.2, $V = 0$ at $z = 0$ and $V = 100$ at $z = d$. Assuming the region between the plates is charge-free,

$$\nabla^2 V = \frac{\partial^2 V}{\partial x^2} + \frac{\partial^2 V}{\partial y^2} + \frac{\partial^2 V}{\partial z^2} = 0 \tag{8.16}$$

With fringing neglected, the potential can vary only with z. Then,

$$\frac{d^2 V}{dx^2} = 0 \tag{8.17}$$

Next, integrating,

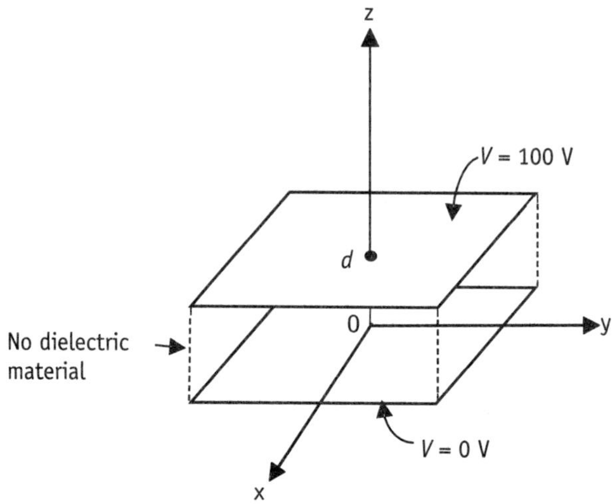

Figure 8.2 Parallel conductors (boundary conditions established).

$$V = Az + B \tag{8.18}$$

The boundary condition $V = 0$ at $z = 0$ requires that $B = 0$. $V = 100$ at $z = d$ gives $A = 100/d$.

Thus,

$$V = 100\left(\frac{z}{d}\right) \quad \text{(V)} \tag{8.19}$$

The electric field intensity, **E**, can now be obtained from:

$$\begin{aligned} \mathbf{E} &= -\nabla V = -\left(\frac{\partial V}{\partial x}\mathbf{a}_x + \frac{\partial V}{\partial y}\mathbf{a}_y + \frac{\partial V}{\partial z}\mathbf{a}_z\right) = -\frac{\partial}{\partial z}\left(100\frac{z}{d}\right)\mathbf{a}_z \\ &= \left(\frac{-100}{d}\right)\mathbf{a}_z \end{aligned} \tag{8.20}$$

Then,

$$\mathbf{D} = \varepsilon\left(\frac{-100}{d}\right)\mathbf{a}_z \quad \left(\text{C/m}^2\right) \tag{8.21}$$

At the conductors,

$$\rho_s = D_s = \pm\varepsilon\left(\frac{100}{d}\right) \quad (\mathrm{C/m^2}) \tag{8.22}$$

where the plus sign applies at $z = d$ and the minus at $z = 0$ [2].

8.7 CARTESIAN PRODUCT SOLUTION

When the potential in Cartesian coordinates varies in more than one direction, Laplace's equation will contain more than one term. Suppose that V is a function of both x and y and has the special form $V = X(x)Y(y)$. Making use of the technique of separation of variables.

$$\frac{\partial^2(XY)}{\partial x^2} + \frac{\partial^2(XY)}{\partial y^2} = 0 \tag{8.23}$$

becomes

$$Y\frac{\partial^2 X}{\partial x^2} + X\frac{\partial^2(Y)}{\partial y^2} = 0 \tag{8.24}$$

or

$$\left(\frac{1}{X}\right)\frac{\partial^2 X}{\partial x^2} + \left(\frac{1}{Y}\right)\frac{\partial^2(Y)}{\partial y^2} = 0 \tag{8.25}$$

Since the first term is independent of y and the second term is independent of x, each may be set equal to a constant. However, the constant for one must be the negative of that for the other. Let the constant be a^2.

$$\left(\frac{1}{X}\right)\frac{\partial^2 X}{\partial x^2} = a^2 \tag{8.26}$$

$$\left(\frac{1}{Y}\right)\frac{\partial^2(Y)}{\partial y^2} = -a^2 \tag{8.27}$$

The general solution for X (for a given a) is:

$$X = A_1 e^{ax} + A_2 e^{-ax} \tag{8.28}$$

or, equivalently,

$$X = A_3 \cosh(ax) + A_4 \sinh(ax) \tag{8.29}$$

The general solution for Y (for a given a) is:

$$Y = B_1 e^{jay} + B_2 e^{-jay} \tag{8.30}$$

or, equivalently,

$$Y = B_3 \cos(ay) = B_4 \sin(ay) \tag{8.31}$$

Therefore, the potential function in the variables x and y can be written:

$$V = X(x)Y(y) = \left(A_1 e^{ax} + A_2 e^{-ax}\right)\left(B_1 e^{jay} + B_2 e^{-jay}\right) \tag{8.32}$$

where

$X(x) = (A_1 e^{ax} + A_2 e^{-ax})$
$Y(y) = (B_1 e^{jay} + B_2 e^{-jay})$

or

$$V = X(x)Y(y) = \left(A_3 \cosh ax + A_4 \sinh ax\right)\left(B_3 \cos ay + B_4 \sin ay\right) \tag{8.33}$$

where

$X(x) = (A_3 \cosh ax + A_4 \sinh ax)$
$Y(y) = (B_3 \cos ay + B_4 \sin ay)$

Because Laplace's equation is a linear, homogeneous equation, a sum of products of the above form is also a solution. Each product corresponds to a different value of a. The most general solution can be generated in this fashion.

Three-dimensional solutions, that is,

$$V = X(x)Y(y)Z(z) \tag{8.34}$$

of similar form to the above equation can be obtained. However, in this case there are two separation constants [4].

8.8 CYLINDRICAL PRODUCT SOLUTION

If a solution of the form,

$$V = R(r)\Phi(\Phi)Z(z) \tag{8.35}$$

is assumed, Laplace's equation becomes:

$$\left(\frac{\Phi Z}{r}\right)\frac{d}{dr}\left(r\frac{dR}{dr}\right)+\left(\frac{RZ}{r^2}\right)\frac{d^2\Phi}{d\Phi^2}+R\Phi\left(\frac{d^2Z}{dz^2}\right)=0 \tag{8.36}$$

Dividing by $R\Phi Z$ and expanding the r-derivative,

$$\left(\frac{1}{R}\right)\left(\frac{d^2R}{dr^2}\right)+\left(\frac{1}{Rr}\right)\left(\frac{dR}{dr}\right)+\left(\frac{1}{r^2\Phi}\right)\left(\frac{d^2\Phi}{d\Phi^2}\right)=-\frac{1}{Z}\left(\frac{d^2Z}{dz^2}\right)=-b^2 \tag{8.37}$$

The r and Φ terms contain no z, and the z term contains neither r nor Φ. They could be set equal to a constant, $-b^2$, as shown above. Then,

$$\frac{1}{Z}\left(\frac{d^2Z}{dz^2}\right)=b^2 \tag{8.38}$$

This equation was encountered in the Cartesian product solution. The solution is:

$$Z = C_1\cosh(bz)+C_2\sinh(bz) \tag{8.39}$$

Now the equation in r and Φ may be further separated as follows:

$$\left(\frac{r^2}{R}\right)\left(\frac{d^2R}{dr^2}\right)+\left(\frac{r}{R}\right)\left(\frac{dR}{dr}\right)+b^2r^2=-\left(\frac{1}{\Phi}\right)\left(\frac{d^2\Phi}{d\phi^2}\right)=a^2 \tag{8.40}$$

The resulting equation in Φ,

$$\left(\frac{1}{\Phi}\right)\left(\frac{d^2\Phi}{d\phi^2}\right)=-a^2 \tag{8.41}$$

has the solution

$$\Phi = C_3\cos(a\phi)+C_4\sin(a\phi) \tag{8.42}$$

The equation in r,

$$\left(\frac{d^2R}{dr^2}\right)+\left(\frac{1}{r}\right)\left(\frac{dR}{dr}\right)+\left[b^2-\left(\frac{a^2}{r^2}\right)\right]R=0 \tag{8.43}$$

is a form of Bessel's differential equation. Its solutions are in the form of power series called Bessel functions [1].

$$R = C_5J_a(br)+C_6N_a(br) \tag{8.44}$$

where

$$J_a(br)=\sum_{m=0}^{\infty}\frac{\left[(-1)^m(br/2)^{(a+2m)}\right]}{\left[m!\Gamma(a+m+1)\right]} \tag{8.45}$$

and

$$N_a(br)=\frac{\left[\cos(a\pi)J_a(br)-J_{-a}(br)\right]}{(\sin a\pi)} \tag{8.46}$$

The series $J_a(br)$ is known as a Bessel function of the *first kind, order a*.

If $a = n$, an integer, the gamma function in the power series may be replaced by $(n + m)!$, and $N_a(br)$ is a Bessel function of the *second kind, order a*. If $a = n$, an integer, $N_a(br)$ is defined as the limit of the above quotient as $a \to n$.

The function $N_a(br)$ behaves like $\ln r$ near $r = 0$ (see Figure 8.3). Therefore, it is not involved in the solution ($C_6 = 0$) whenever the potential is known to be finite at $r = 0$.

For integral order n and large argument x, the Bessel functions behave like damped sine waves:

$$J_n(x) \sim \left(\frac{2}{\pi x}\right)^{1/2} \cos\left[x - \left(\frac{\pi}{4}\right) - \left(\frac{n\pi}{2}\right)\right] \tag{8.47}$$

$$N_n(x) \sim \left(\frac{2}{\pi x}\right)^{1/2} \sin\left[x - \left(\frac{\pi}{4}\right) - \left(\frac{n\pi}{2}\right)\right] \tag{8.48}$$

See Figure 8.3.

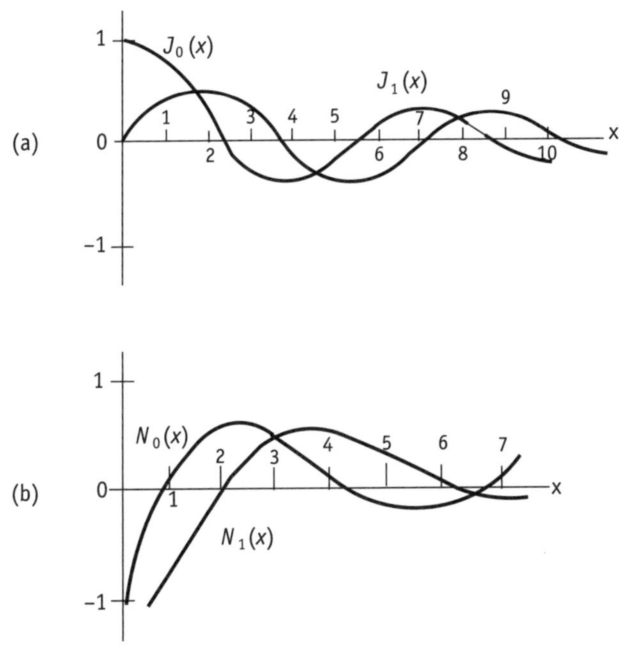

Figure 8.3 Bessel function (a) first kind, *n* order, (b) second kind, *n* order.

8.9 SPHERICAL PRODUCT SOLUTION

Of particular interest are those problems in spherical coordinates in which V may vary with r and θ but not with ϕ. For a product solution,

$$V = R(r)\Theta(\theta) \tag{8.49}$$

Laplace's equation becomes:

$$\left[\left(\frac{r^2}{R}\right)\left(\frac{d^2R}{dr^2}\right)+\left(\frac{2r}{R}\right)\left(\frac{dR}{dt}\right)\right]+\left[\left(\frac{1}{\Theta}\right)\left(\frac{d^2\Theta}{d\theta^2}\right)+\left\{\left[\frac{1}{(\Theta\tan\theta)}\right]\left(\frac{d\Theta}{d\theta}\right)\right\}=0 \tag{8.50}$$

The separation constant is chosen as $n(n + 1)$, where n is an integer (for reasons that will become apparent). The two separated equations are:

$$r^2\left(\frac{d^2R}{dr^2}\right)+2r\left(\frac{dR}{dr}\right)-n(n+1)R=0 \tag{8.51}$$

and

$$\left(\frac{d^2\Theta}{d\theta^2}\right)+\left(\frac{1}{\tan\theta}\right)\left(\frac{d\Theta}{d\theta}\right)\Bigg\}+n(n+1)\Theta=0 \tag{8.52}$$

The equation in r has the solution:

$$R = C_1 r^n + C_2 r^{-(n+1)} \tag{8.53}$$

The equation in θ possesses (unlike Bessel's equation) a polynomial solution of degree n in the variable $\xi = \cos\theta$, given by:

$$P_n(\xi) = \left(\frac{1}{2^n n!}\right)\left[\frac{d^n}{d\xi^n}(\xi^2 - 1)^n\right] \tag{8.54}$$

where $n = 0, 1, 2, \ldots$

The polynomial $P_n(\xi)$ is the *Legendre polynomial of order* n. There is a second, independent solution, $Q_n(\xi)$, which is logarithmically infinite at $\xi = \pm 1$ (i.e., $\theta = 0, \pi$) [4].

8.10 PROBLEMS AND SOLUTIONS: LAPLACE'S EQUATION

Problem 8.1

The potential has the value:

- V_1 on $1/n$ of the circle;
- 0 on the rest of the circle.

The entire region inside the circle is charge free (see Figure 8.4).

Find the potential at the center of the circle.

Problem 8.2

Show how mean value theorem follows the result of the previously solved problem.

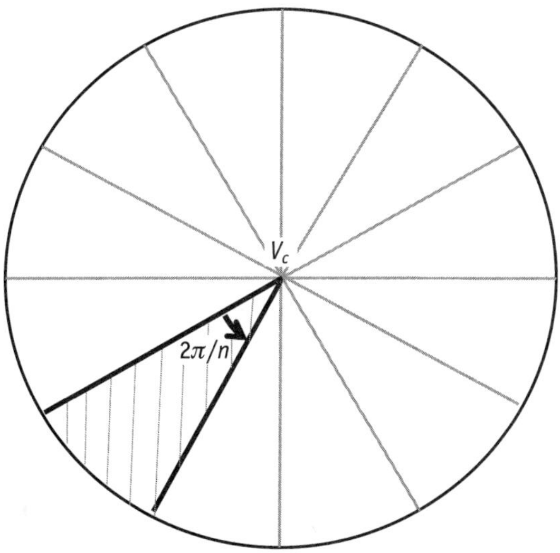

Figure 8.4 Circle highlighting a specific single sector.

Problem 8.3

Prove that with a charge-free region then potential cannot attain a maximum value.

Problem 8.4

Find the potential function for the region between the parallel circular disks of Figure 8.5. Neglect fringing.

Problem 8.5

Two parallel conducting plates in free space are at

$$y = 0 \quad \text{and} \quad y = 0.02\text{m}.$$

The zero-voltage reference is at $y = 0.01$m. If:

$$\mathbf{D} = 253\mathbf{y}\ (\text{nC/m}^2)$$

between the conductors, determine conductor voltages.

Problem 8.6

The parallel conducting disks in Figure 8.6 are separated by 5 mm and contain a dielectric for which $e_r = 2.2$.

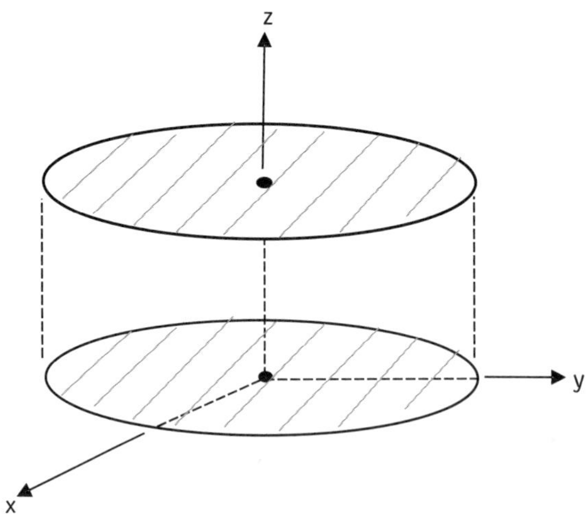

Figure 8.5 Circle divided into sectors.

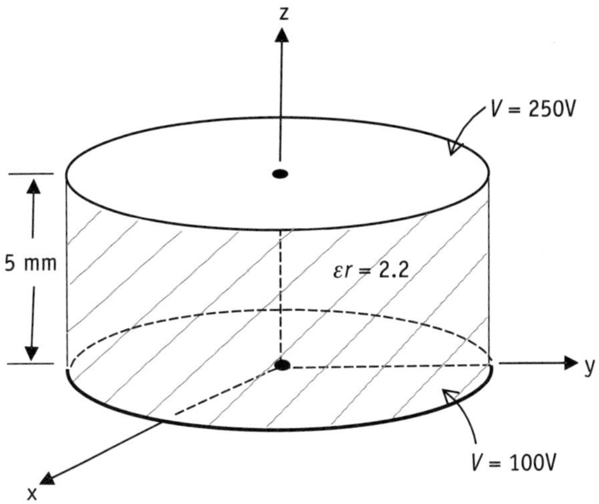

Figure 8.6 Circular disks separated by dielectric.

Determine charge densities on the disks.

References

[1] Boyce, W. E., *Calculus*, New York: John Wiley & Sons, 1988.

[2] Shadowitz, A., *The Electromagnetic Field*, New York: Dover Publications, 1975.

[3] Warnick, K. F., and P. Russer, *Problem Solving in Electromagnetics, Microwave Circuit and Antenna Design for Communications Engineering*, Norwood, MA: Artech House, 2006.

[4] Russer, P., *Electromagnetics, Microwave Circuit and Antenna Design for Communications Engineering*, Norwood, MA: Artech House, 2006.

PART 2

MAGNETOSTATICS: BASIC PRINCIPLES

9

AMPERE'S LAW AND THE MAGNETIC FIELD

9.1 MAGNETOSTATICS

A static magnetic field can originate from either a constant current source or a permanent magnet. This chapter is devoted to the discussion of nontime-varying magnetic fields that are produced from constant current sources. Time-variable magnetic fields, that is, those which coexist with time-variable electric fields, will be examined in Volume 2 of this series, beginning with Chapter 12.

9.2 BIOT-SAVART LAW

A differential *magnetic field strength*, *dH*, results from a differential current element $I\,d\mathbf{l}$. The field varies inversely with distance squared. It is also independent of the surrounding medium and has a direction given by the cross-product of $I\,d\mathbf{l}$ and $\mathbf{a}_r$ (or $\mathbf{r}$) [1]. This relationship is known as the Biot-Savart law:

$$d\mathbf{H} = Id\mathbf{l} \times \frac{\mathbf{a}_r}{4\pi R^2} = Id\mathbf{l} \times \frac{r}{4\pi R^2} \quad \text{(A/m)} \tag{9.1}$$

The direction of **R** must be from the current element to the point at which $d\mathbf{H}$ is to be determined. This is shown in Figure 9.1.

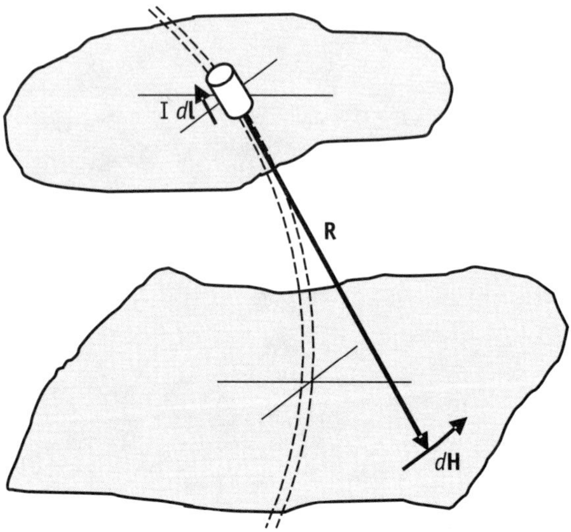

Figure 9.1 Generalized illustration of the Biot-Savart law.

Current elements have no separate existence. All elements making up the complete current filament contribute to **H** and, therefore, must be included. As a result, the summation leads to the integral form of the Biot-Savart law:

$$\mathbf{H}\oint I\,d\mathbf{l} \times \frac{\mathbf{a}_r}{4\pi R^2} = \oint I\,d\mathbf{l} \times \frac{r}{4\pi R^2} \quad \text{(A/m)} \tag{9.2}$$

The closed line integral simply requires that all I current elements be included in order to obtain the complete **H** (the contour may close at ∞).

Example 9.1

An infinitely long, straight, filamentary current I along the z-axis in cylindrical coordinates is shown in Figure 9.2 [2].

A point in the $z = 0$ plane is selected with no loss in generality. In differential form, the equation may be expressed as:

$$d\mathbf{H} = I\,dz\,\mathbf{z} \times \frac{(r\mathbf{r} - z\mathbf{z})}{4\pi\left(r^2 + z^2\right)^{3/2}}$$

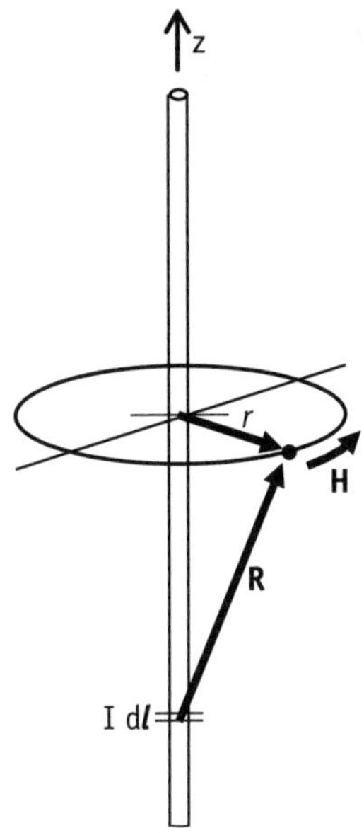

Figure 9.2 Biot-Savart law illustrated in cylindrical coordinates.

The variable of integration is z. Since $\mathbf{a}_\phi$ does not change with z, it may be removed from the integrand before integrating [3]. Hence,

$$\mathbf{H} = \left[\int_{-\infty}^{\infty} \frac{I\ r\ dz}{4\pi\left(r^2 + z^2\right)^{\frac{3}{2}}}\right]\mathbf{a}_\phi = \left(\frac{I}{2\pi r}\right)\mathbf{a}_\phi$$

This important result shows that **H** is inversely proportional to the radial distance, and the direction is in agreement with the right-hand rule; that is, where the four curled fingers of the right-hand point in the direction of the H-field when the z-axis aligned conductor is grasped by the hand such that the thumb points in the direction of the current (see Figure 9.2).

Continuing, the magnetic fields of sheet currents and volume currents are also given by the integral form of the Biot-Savart law. However, $I\,d\mathbf{l}$ is replaced by $\mathbf{K}dS$ (sheet current) or $\mathbf{J}dv$ (volume current) and with the integration extended, respectively, over the entire sheet or volume.

A particular case of importance is the infinite plane sheet of constant density, **K**. The field in this case is constant.

$$\mathbf{H} = \frac{1}{2}\mathbf{K} \times \mathbf{a}_n = \frac{1}{2}\mathbf{K} \times \mathbf{n} \quad \text{(A/m)} \tag{9.3}$$

where $\mathbf{n} = \mathbf{a}_n$.

9.3 AMPERE'S LAW

The line integral of the tangential component of **H** around a closed path is equal to the current enclosed by the path:

$$\oint \mathbf{H} \cdot d\mathbf{l} = I_{\text{enclosed}} \tag{9.4}$$

This is Ampere's law [1].

At first glance, a person would think that the law is used to determine the current I by integration. Instead, in practical cases, the current is usually known. The law provides a method of finding **H**. It is quite similar to the use of Gauss' law to find **D** given a known charge distribution.

In order to use Ampere's law to determine **H** there must be a considerable degree of symmetry in the problem. As such, two conditions must be met:

1. At each point of the closed path, **H** is either tangential or normal to the path;
2. H has the same value at all points of the path where **H** is tangential.

Thus, the Biot-Savart law can be used to help select a path which meets the two above-stated conditions [4]. In most cases, a proper path will be evident.

9.4 CURL

The curl of a vector field A results in another vector field. Point P in Figure 9.3 lies in the plane area ΔS bounded by the closed curve C.

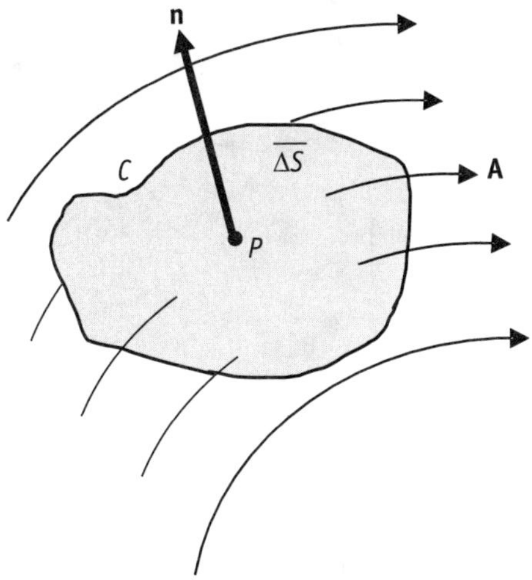

Figure 9.3 Illustration of the curl.

In the integration that defines the curl, C is traversed such that the enclosed area is as shown in the figure. The unit normal, **n**, (also sometimes denoted as $\mathbf{a}_n$ in vector mathematics), is determined by the right-hand rule (as shown in Figure 9.3). Then, the component of the curl of A in the direction **n**, is defined as [3]:

$$(\text{curl } \mathbf{A}) \cdot \mathbf{n} = \lim_{\Delta S \to 0} \frac{\left(\oint \mathbf{A} \cdot d\mathbf{l}\right)}{\Delta S} \tag{9.5}$$

In the coordinate systems, curl **A** is completely specified by its components along the three unit vectors. The x-component in Cartesian coordinates is defined by the taking as the contour C a square in the x = constant plane through P, as shown in Figure 9.4.

$$(\text{curl } \mathbf{A}) \cdot \mathbf{n}_x = \lim_{\Delta y \Delta x \to 0} \frac{\left(\oint \mathbf{A} \cdot d\mathbf{l}\right)}{\Delta y \Delta z} \tag{9.6}$$

If $\mathbf{A} = A_x\mathbf{a_x} + A_y\mathbf{a_y} + A_z\mathbf{a_z}$ at the corner of ΔS closest to the origin (point #1, see Figure 9.4), then,

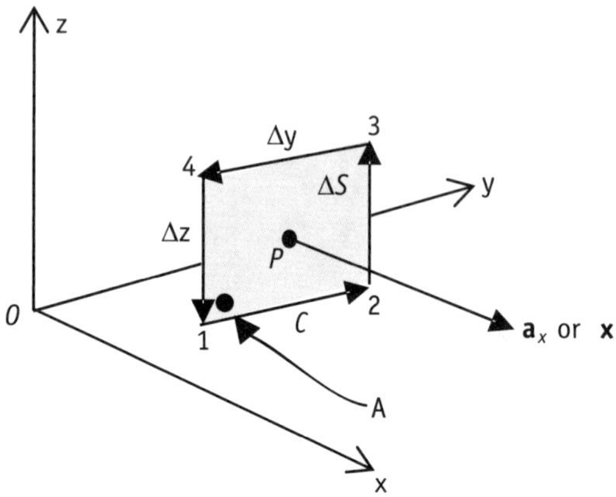

Figure 9.4 Curl in three components.

$$\oint = \int_1^2 + \int_2^3 + \int_3^4 + \int_4^1 = A_y \Delta_y + \left[A_x + \left(\frac{\partial A_z}{\partial y} \right) \Delta y \right] \Delta z$$
$$+ \left[A_y + \left(\frac{\partial A_y}{\partial z} \right) \Delta z \right] (-\Delta y) + A_x (-\Delta z) \tag{9.7}$$
$$= \left(\frac{\partial A_z}{\partial y} - \frac{\partial A_y}{\partial z} \right) \Delta y \Delta z$$

and,

$$(\text{curl } \mathbf{A}) \cdot \mathbf{a_x} = \left(\frac{\partial A_z}{\partial y} - \frac{\partial A_y}{\partial z} \right) \tag{9.8}$$

The y- and z-components can be determined in a similar fashion. Combining the three components:

Cartesian:

$$\text{curl } \mathbf{A} = \left(\frac{\partial A_z}{\partial y} - \frac{\partial A_y}{\partial z} \right) \mathbf{a_x} + \left(\frac{\partial A_x}{\partial z} - \frac{\partial A_z}{\partial x} \right) \mathbf{a_y} + \left(\frac{\partial A_y}{\partial x} - \frac{\partial A_x}{\partial y} \right) \mathbf{a_z} \tag{9.9}$$

A third-order determinant can be written. The expansion gives the Cartesian curl of **A**.

$$\text{curl } \mathbf{A} = \begin{vmatrix} \mathbf{a_x} & \mathbf{a_y} & \mathbf{a_z} \\ \partial/\partial x & \partial/\partial y & \partial/\partial z \\ A_x & A_y & A_z \end{vmatrix} \tag{9.10}$$

The elements in the second row are the del operator components (see Section 1.2). This suggests that $\nabla \times \mathbf{A}$ can be written for curl **A**. As with other expressions from vector analysis, this convenient notation is used for curl **A** in order coordinate systems even though ∇ is defined only in Cartesian coordinates.

Expressions for curl **A** in cylindrical and spherical coordinates can be derived in the same manner as above, though with more difficulty. See the following:

Cylindrical:

$$\begin{aligned} \text{curl } \mathbf{A} &= \left[\frac{1}{r}\frac{\partial A_z}{\partial \phi} - \frac{\partial A_\phi}{\partial z}\right]\mathbf{a_r} + \left(\frac{\partial A_r}{\partial z} - \frac{\partial A_z}{\partial r}\right)\mathbf{a_\phi} \\ &+ \frac{1}{r}\left[\partial\left(\frac{rA_\theta}{\partial r} - \frac{\partial A_r}{\partial \phi}\right)\right]\mathbf{a_z} \end{aligned} \tag{9.11}$$

Spherical:

$$\begin{aligned} \text{curl } \mathbf{A} &= \frac{1}{(r\sin\theta)}\left[\frac{\partial(A_\phi \sin\theta)}{\partial\theta} - \frac{\partial A_\theta}{\partial \phi}\right]\mathbf{a_r} \\ &+ \left(\frac{1}{r}\right)\left[\left(\frac{1}{\sin\theta}\right)\left(\frac{\partial A_r}{\partial \phi}\right) - \frac{\partial(rA_\phi)}{\partial r}\right]\mathbf{a_\theta} \\ &+ \frac{1}{r}\left[\frac{\partial(rA_\theta)}{\partial r} - \frac{\partial A_r}{\partial \theta}\right)\mathbf{a_\phi} \end{aligned} \tag{9.12}$$

Two properties of the curl operator are frequently useful:

1. *The divergence of a curl is zero*; that is,

$$\nabla \cdot (\nabla \times \mathbf{A}) = 0 \tag{9.13}$$

for any vector field A;

2. *The curl of a gradient is zero*; that is,

$$\nabla \cdot (\nabla f) = 0 \tag{9.14}$$

for any scalar function of position f [5].

Under static conditions, $\mathbf{E} = -\nabla V$, and so from above property (2),

$$\nabla \times \mathbf{E} = 0 \tag{9.15}$$

9.5 CURRENT DENSITY J AND $\nabla \times \mathbf{H}$

The x-component of $\nabla \times \mathbf{H}$ is determined by $\oint \mathbf{H} \cdot d\mathbf{l}$, where the path lies in a plane normal to the x-axis. Ampere's law states that this integral is equal to the current enclosed. The direction is $\mathbf{a_x}$ so the current can be called I_x [1]. Thus,

$$(\text{curl } \mathbf{A}) \cdot \mathbf{a_x} = \lim_{\Delta S \to 0} \frac{I_x}{\Delta S} = J_x \tag{9.16}$$

The x-component of the current density $\mathbf{J}$. Likewise for the y and z directions. Consequently,

$$\nabla \times \mathbf{H} = \mathbf{J} \tag{9.17}$$

This important result is one of Maxwell's equations for static fields. If $\mathbf{H}$ is known throughout a particular region, then $\nabla \times \mathbf{H}$ will produce $\mathbf{J}$ for that region.

9.6 MAGNETIC FLUX DENSITY B

Like $\mathbf{D}$, the magnetic field strength $\mathbf{H}$ depends only on (moving) charges and is independent of the medium. The force field associated with $\mathbf{H}$ is the *magnetic flux density* $\mathbf{B}$, which is given by

$$\mathbf{B} = \mu \mathbf{H} \tag{9.18}$$

where: $\mu = \mu_0\mu_r$ is the *permeability* of the medium. The unit of **B** is the *tesla*,

$$1\text{T} = 1\ \text{N/Am} \tag{9.19}$$

The free-space permeability μ_0 has the numerical value of $4\pi \times 10^{-7}$ and has the units henries per meter, H/m; μ_r, the *relative permeability* of the medium, is a pure number very near to unity except for a small group of ferromagnetic materials which will be treated in Chapter 11.

Magnetic flux, Φ, through a surface is defined as:

$$\Phi = \int_S \mathbf{B} \cdot d\mathbf{S} \tag{9.20}$$

The sign on Φ may be positive or negative depending upon the choice of the surface normal in $d\mathbf{S}$. The unit of magnetic flux is the weber, Wb [4]. The various magnetic units are related by:

$$1\text{T} = 1\ \text{Wb/m}^2 \tag{9.21}$$

$$1\text{H} = 1\ \text{Wb}/A \tag{9.22}$$

Example 9.2

Find the flux crossing the portion of the plane $\phi = \pi/4$ defined by $0.01\text{m} < r < 0.05\text{m}$ and $0 < z < 2\text{m}$ (see Figure 9.5). A current filament carrying 2.50A of current along the z-axis is in the $\mathbf{a}_z$ or z direction.

Solution: See the following equations:

$$\mathbf{B} = \mu_0 \mathbf{H} = \mu_0 \left(\frac{I}{2\pi r}\right) \boldsymbol{\phi}_\phi = \mu_0 \left(\frac{I}{2\pi r}\right) \boldsymbol{\phi}$$

$$d\mathbf{S} = dr\,dz\,\mathbf{a}_\phi = dr\,dz\,\boldsymbol{\phi}$$

$$\boldsymbol{\Phi} = \int_0^2 \int_{0.01}^{0.05} \left(\frac{\mu_0 I}{2\pi r}\right) \boldsymbol{\phi} \cdot (dr\,dz))\boldsymbol{\phi} = \left(\frac{2\mu_0 I}{2\pi}\right) \ln\left(\frac{0.05}{0.01}\right)$$

$$= 1.61 \times 10^{-6}\ \text{Wb or } 1.61\ \mu\text{Wb}$$

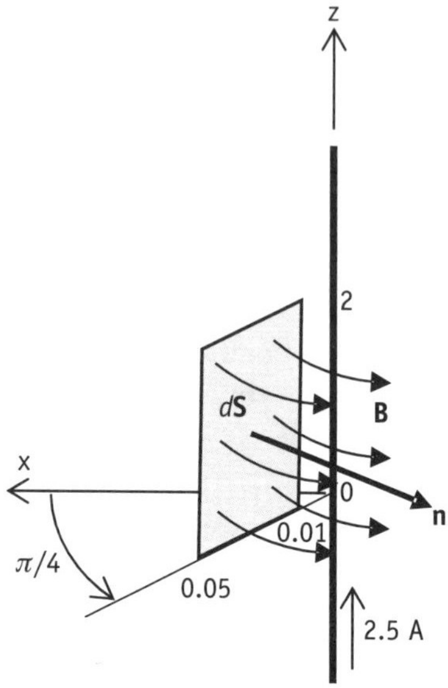

Figure 9.5 Magnetic flux crossing a portion of a plane.

It should be observed that the lines of magnetic flux Φ are closed curves with no starting or termination points. This is in contrast with electric flux ψ which originates on positive charge and terminates on negative charge. In Figure 9.6, all of the magnetic flux Φ that enters the closed surface must leave the surface.

Thus, **B** fields have no sources or sinks which is mathematically expressed by:

$$\nabla \cdot \mathbf{B} = 0 \tag{9.23}$$

9.7 VECTOR MAGNETIC POTENTIAL A

Electric field intensity **E** was first obtained from known charge configurations. Later, electric potential V was developed, and it was found that **E** could be obtained as the *negative gradient of* V; that is, $\mathbf{E} = -\nabla V$. Laplace's equation provided a method of obtaining V from known potentials on the boundary conductors. Similarly, a *vector magnetic potential*, **A**, is defined such that

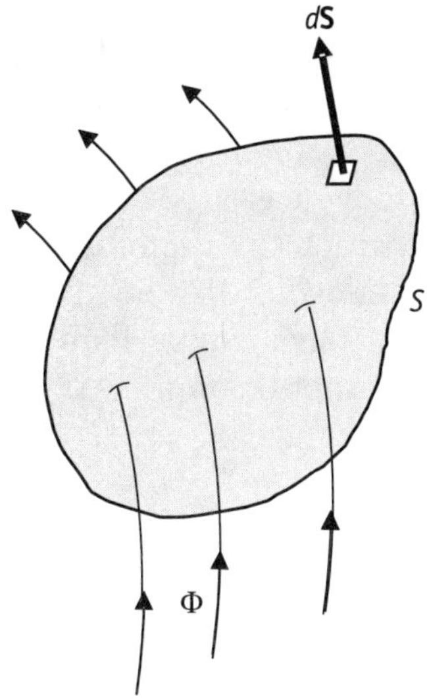

Figure 9.6 Magnetic flux entering the closed surface must exit the surface.

$$\nabla \times \mathbf{A} = \mathbf{B} \tag{9.24}$$

It serves as an intermediate quantity from which **B**, and, hence, **H** can be calculated. Note that the definition of **A** is consistent with the requirement that $\nabla \cdot \mathbf{B} = 0$. The units of **A** are Wb/m or T · m.

If the additional condition:

$$\nabla \cdot \mathbf{A} = 0 \tag{9.25}$$

is imposed, then the vector magnetic potential **A** can be determined form the known currents in the region of interest. For the three standard current configurations, the expressions are:

$$\text{Current filament:} \quad \mathbf{A} = \oint \left(\frac{\mu I}{4\pi r} \right) d\mathbf{I} \tag{9.26}$$

$$\text{Sheet current:} \quad \mathbf{A} = \int_S \left(\frac{\mu \mathbf{K}}{4\pi R} \right) dS \tag{9.27}$$

$$\text{Volume current:} \quad \mathbf{A} = \int_V \left(\frac{\mu \mathbf{J}}{4\pi R} \right) dv \tag{9.28}$$

R in each of the above cases is the distance from the current element to the point at which the vector magnetic potential is being calculated. Like the analogous integral for the electric potential (review Section 5.4), the above expressions for **A** presuppose a zero level at infinity. They cannot be applied if the current distribution itself extends to infinity [4].

Example 9.3

Investigate the vector magnetic potential for the infinite, straight, current filament I in free space. In Figure 9.7, the current filament is along the z-axis. The observation point is (x, y, z).

The particular current element

$$I\,d\mathbf{l} = I\,dl\,\mathbf{a}_z = I\,dl\,\mathbf{z}$$

at $l = 0$ is shown, where l is the running variable along the z-axis.

It is clear that the integral

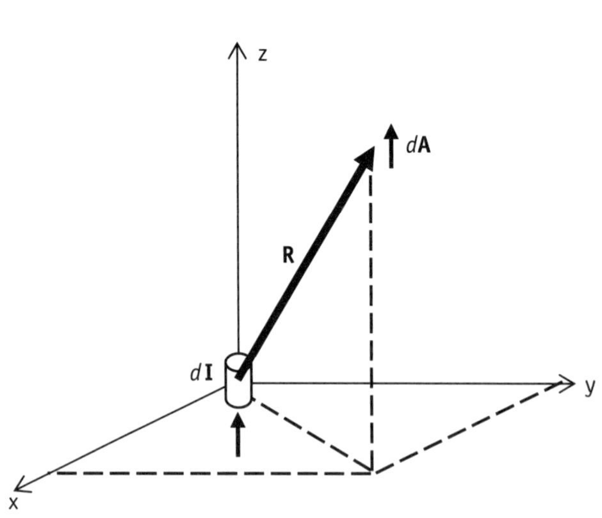

Figure 9.7 Vector magnetic potential illustration.

$$\mathbf{A} = \int_{-\infty}^{+\infty} \frac{(\mu_0 I\, dl))}{(4\pi R)} \mathbf{z}$$

does *not* exist since, when l is large, $\mathbf{R} \simeq l$. This is a case of a current distribution that extends to infinity.

However, it is possible to consider the *differential* vector potential:

$$d\mathbf{A} = \left\{ \frac{(\mu_0 I\, dl)}{\left[4\pi\left(x^2 + y^2 + z^2\right)^{1/2}\right]} \right\} \mathbf{z}$$

and

$$d\mathbf{B} = \nabla \times d\mathbf{A}$$
$$= \left(\frac{\mu_0 I\, dl}{4\pi}\right) \left[\left[\frac{-y}{\left(x^2 + y^2 + z^2\right)^{3/2}} \right] \mathbf{x} + \left[\frac{x}{\left(x^2 + y^2 + z^2\right)^{3/2}} \right] \mathbf{y} \right]$$

This result agrees with that for $d\mathbf{H} = (1/\mu_0)\, d\mathbf{B}$ as given by the Biot-Savart law.

9.8 STOKES' THEOREM

Consider an open surface S in which the boundary is a closed curve C. *Stokes' theorem* states that the integral of the tangential component of a vector field **F** around C is equal to the integral of the normal component of curl **F** over S [3].

$$\oint \mathbf{F} \cdot d\mathbf{l} = \int_S (\nabla \times \mathbf{F}) \cdot d\mathbf{S} \tag{9.29}$$

If **F** is chosen to be the vector magnetic potential **A**, Stokes' theorem gives:

$$\oint \mathbf{A} \cdot d\mathbf{l} = \int_S \mathbf{B} \cdot d\mathbf{S} = \Phi \tag{9.30}$$

9.9 PROBLEMS AND SOLUTIONS

Problem 9.1

An infinitely long, straight, filamentary current, I, runs along the z-axis as shown in Figure 9.8. In cylindrical coordinates, derive the equation for the H-field as a function of the current and the distance, r, away from the filament.

Problem 9.2

In this problem, use Ampere's law to obtain **H** for the same figure (Figure 9.8) of the previous problem.

Problem 9.3

Find the flux crossing the portion of the plane $\phi = \pi/4$ defined by $0.01\text{m} < r < 0.05\text{m}$ and $0\text{m} < z < 2\text{m}$. See Figure 9.9. A current filament of 2.5A along the z-axis is in the z direction.

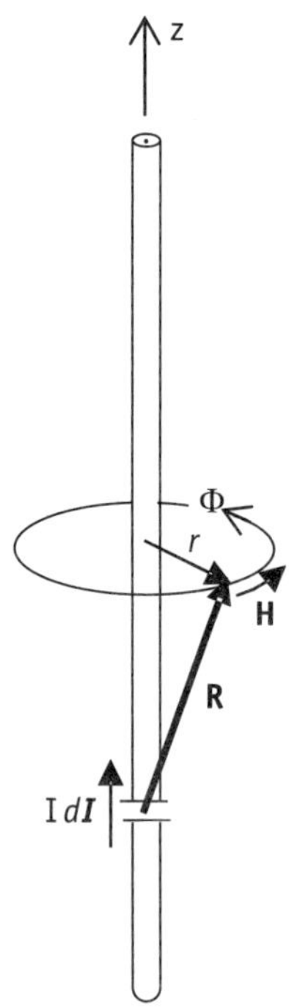

Figure 9.8 Biot-Savart law (i.e., H-field created from induced current, I).

Problem 9.4

Perform an investigation of the vector magnetic potential for the infinite, straight current filament, I, in free space (see Figure 9.10).

Problem 9.5

Find **H** at the center of the square current loop having sides of length L (see Figure 9.11).

Problem 9.6

A current filament of 5.0A in the y direction is parallel to the y-axis at $x = 2\text{m}$, $z = -2\text{m}$. Find **H** at the origin (see Figure 9.12).

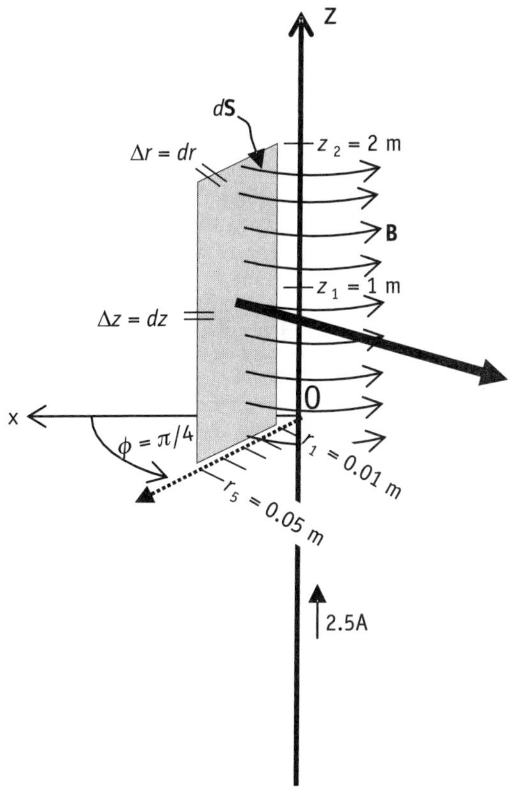

Figure 9.9 Ampere's law (i.e., current created from induced H-field).

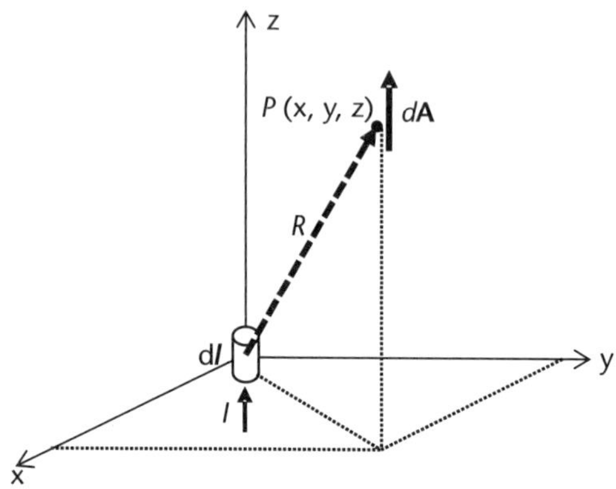

Figure 9.10 Illustration for vector magnetic potential investigation.

Problem 9.7

Determine an expression for **H** that is due to an infinite plane current sheet of uniform density, **K** (see Figure 9.13).

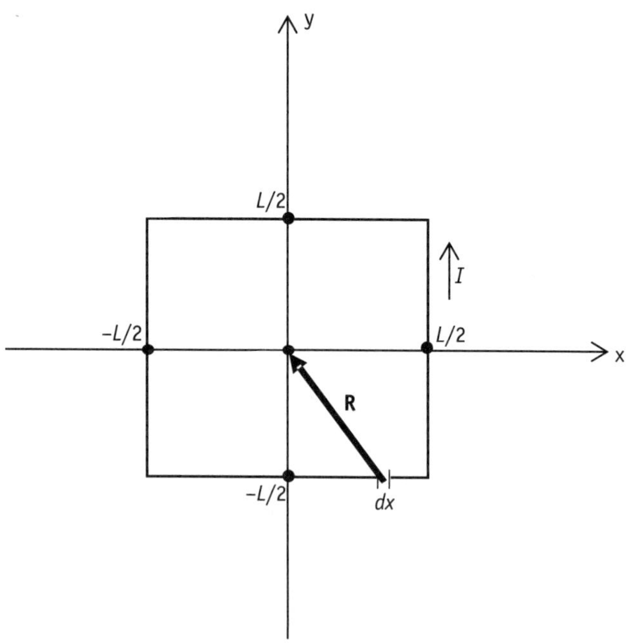

Figure 9.11 Illustration for Problem 9.5.

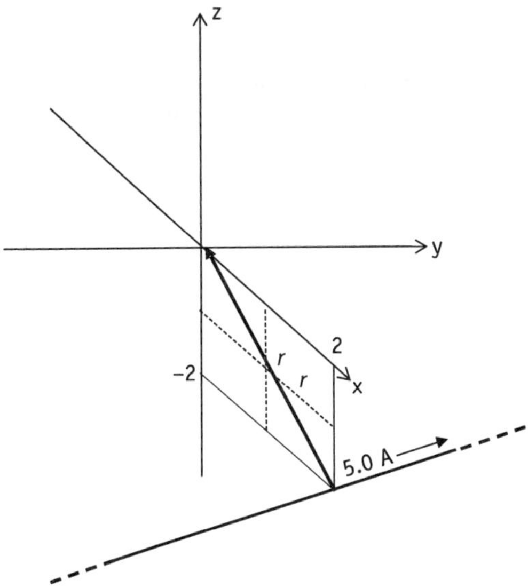

Figure 9.12 Illustration for Problem 9.6.

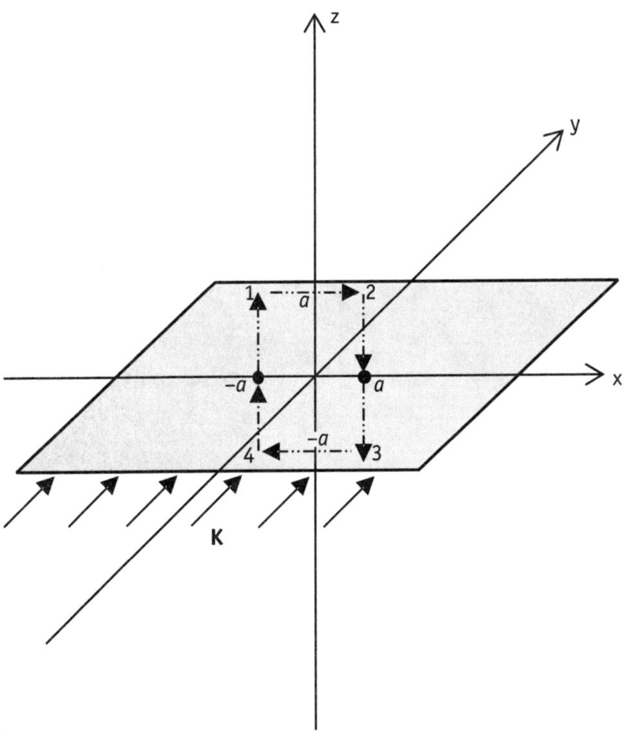

Figure 9.13 Illustration for Problem 9.7.

Problem 9.8

A current sheet, $K = 10\mathbf{z}$ (A/m), lies in the $x = 5$m plane. A second sheet, $\mathbf{K} = -10\mathbf{z}$ (A/m), lies in the $x = -5$m plane. Find **H** at all points (see Figure 9.14).

Problem 9.9

A thin hollow cylinder conductor of radius a and infinite length carries a current, **I**. Find H at all points using Ampere's law (see Figure 9.15).

Problem 9.10

Determine **H** for a solid cylindrical conductor of radius a where the current, **I**, is uniformly distributed over the cross section (see Figure 9.16).

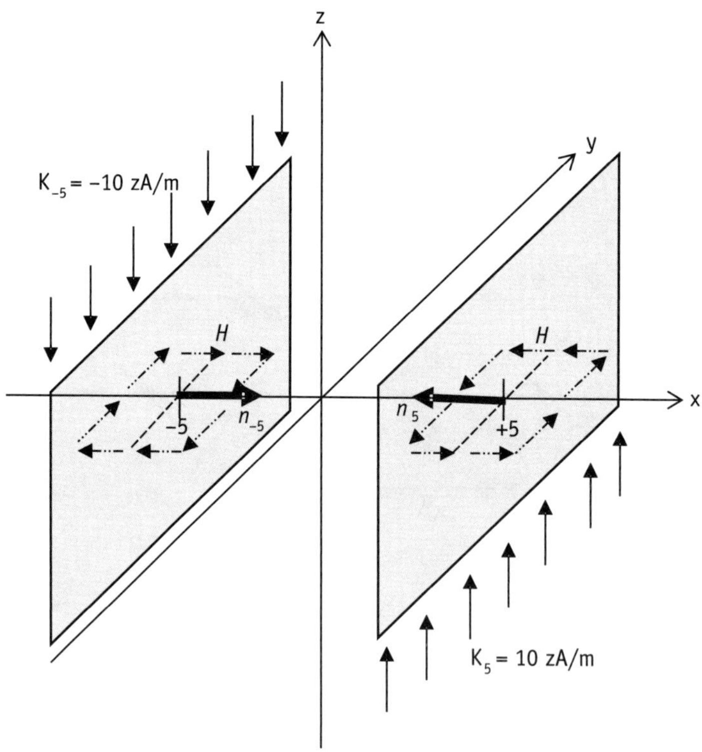

Figure 9.14 Illustration for Problem 9.8.

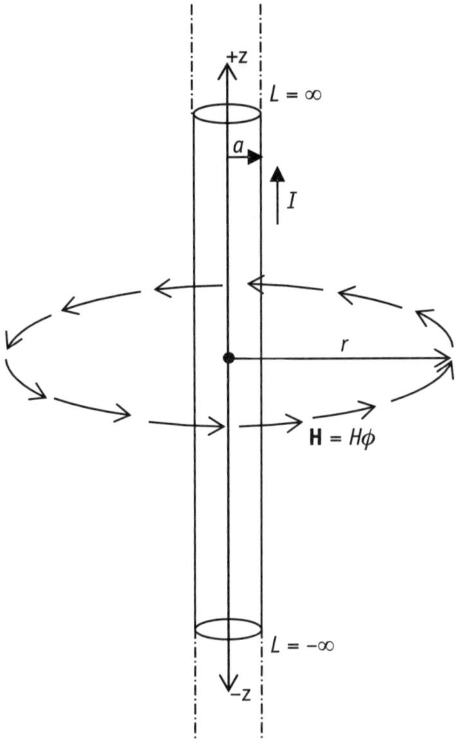

Figure 9.15 Illustration for Problem 9.9.

Problem 9.11

In the region $0 < r < 0.5$m in cylindrical coordinates, the current density is

$\mathbf{J} = 4.5\ e^{-2r}\mathbf{z}$

and

$\mathbf{J} = 0\ \mathbf{z}$ elsewhere.

Use Ampere's law to find **H** (see Figure 9.17).

Problem 9.12

A current sheet, $\mathbf{K} = 6.0\mathbf{x}$ (A/m), lies in the $z = 0$ plane with a current located $y = 0$, $z = 4$m shown in Figure 9.18.

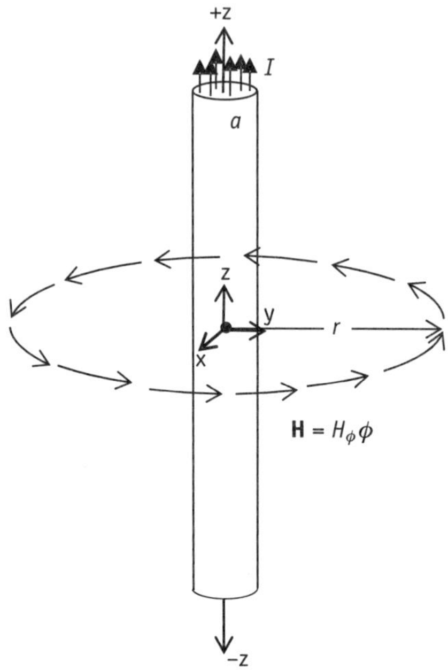

Figure 9.16 Illustration for Problem 9.10.

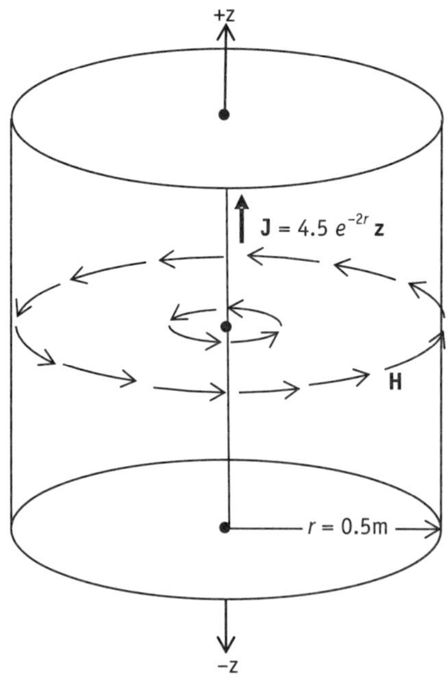

Figure 9.17 Illustration for Problem 9.11.

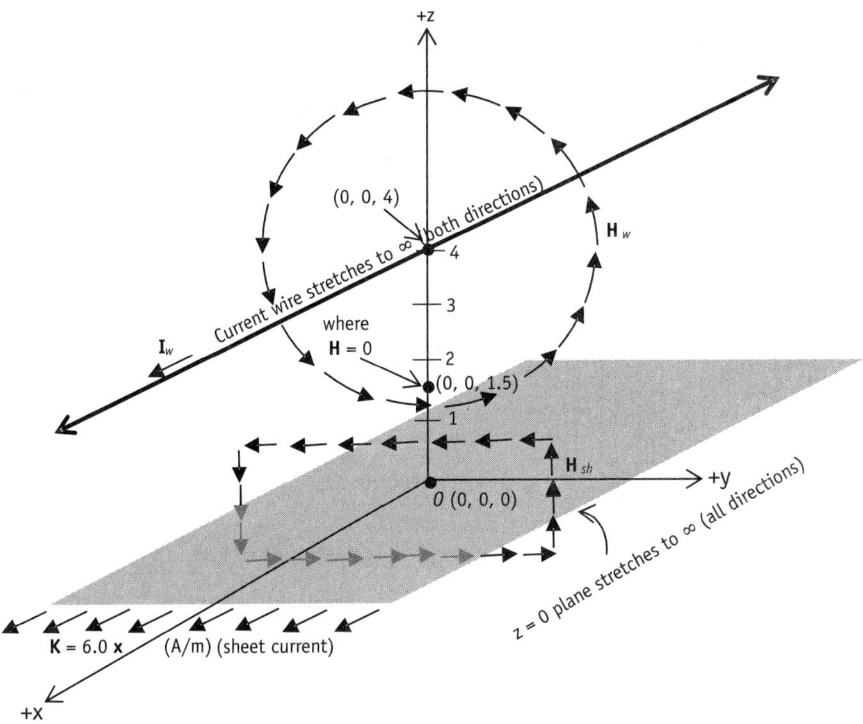

Figure 9.18 Illustration for Problem 9.12.

Determine I_w and its direction of $\mathbf{H} = 0$ at point (0, 0, 1.5m).

References

[1] Shadowitz, A., *The Electromagnetic Field*, New York: Dover Publications, Inc., 1975.

[2] Warnick, K. F., and P. Russer, *Problem Solving in Electromagnetics, Microwave Circuit and Antenna Design for Communications Engineering*, Norwood, MA: Artech House, 2006.

[3] Boyce, W. E., *Calculus*, New York: John Wiley & Sons, 1988.

[4] Russer, P., *Electromagnetics, Microwave Circuit and Antenna Design for Communications Engineering*, Norwood, MA: Artech House, 2006.

[5] Cheng, D. K., *Fundamentals of Engineering Electromagnetics*, Reading, MA: Addison-Wesley Publishing Company, Inc., 1993.

10

FORCES AND TORQUES IN MAGNETIC FIELDS

10.1 MAGNETIC FORCE ON PARTICLES

A charged particle *in motion* in a magnetic field experiences a force at right angles to its velocity—with a magnitude proportional to the charge—the velocity and the magnetic flux density. The complete expression is given by the cross-product [1].

$$\mathbf{F} = Q\mathbf{U} \times \mathbf{B} \tag{10.1}$$

Therefore, the direction of a particle in motion can be changed by a magnetic field. The magnitude of the velocity, U, and, consequently, the kinetic energy, will remain the same. This is in contrast to the electric field where the force

$$\mathbf{F} = Q\mathbf{E} \tag{10.2}$$

does work on the particle and, for this reason, changes its kinetic energy [1].

If the field $\mathbf{B}$ is uniform throughout a region and the particle has an initial velocity normal to the field, the path of the particle is a circle of a certain radius, r. The force of the field is of magnitude

$$F = |Q|UB \tag{10.3}$$

and is directed toward the center of the circle. The centripetal acceleration is of magnitude

$$\omega^2 r = \frac{U^2}{r} \tag{10.4}$$

Then, by Newton's second law:

$$|Q|UB = m\left(\frac{U^2}{r}\right) \tag{10.5}$$

or

$$r = \frac{mU}{|Q|B} \tag{10.6}$$

Observe that r is a measure of the particle's linear momentum, mU [2]. See Examples 10.1 and 10.2.

Example 10.1

Find the force on a particle of mass 1.70×10^{-17} kg and charge 1.60×10^{-19} C if it enters a field $\mathbf{B} = 5$ mT with an initial speed of 83.5 km/s.

Solution: See Figure 10.1. Unless directions are known for **B** and $\mathbf{U}_o$, the particle's initial velocity, the force cannot be calculated. Therefore, under the assumption that **B** and $\mathbf{U}_o$ are perpendicular to each other, the following may be performed:

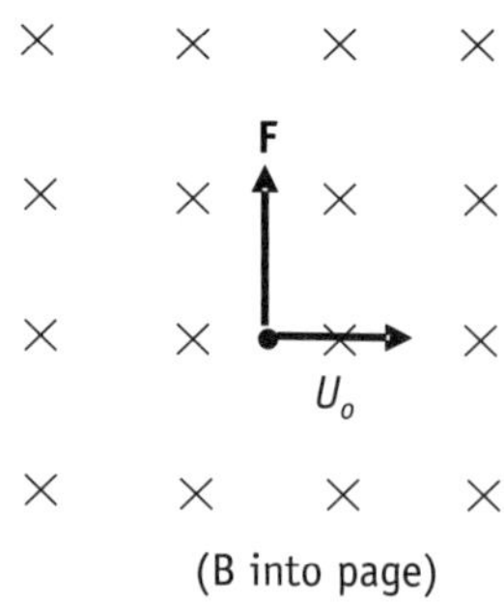

Figure 10.1 Magnetic force illustration for Example 10.1.

$$F = |Q|UB = |1.60 \times 10^{-19}\ \text{C}|(83.5 \times 10^{3}\ \text{m/s})(5 \times 10^{-3}\ \text{T})$$
$$= 6.68 \times 10^{-17}\ \text{N}$$

where

$Q = 1.60 \times 10^{-19}$ C

$U = 83.5 \times 10^{3}$ m/3

$B = 5$ mT

Example 10.2

For the Example 10.1 particle, find the radius of the circular path and the time required for one revolution.

$$r = \frac{mU}{|Q|B} = \frac{(1.70 \times 10^{-17}\ \text{kg})(83.5 \times 10^{3}\ \text{m/s})}{(1.60 \times 10^{-19}\ \text{C})(5 \times 10^{-3}\ \text{T})}$$
$$= 0.1774\ \text{m}$$

where

$m = 1.70 \times 10^{-17}$ kg

$U = 83.5 \times 103$ m/s

$Q = 1.60 \times 10^{-19}$ C

$B = 5$ mT

$$T = \frac{2\pi r}{U} = 2\pi\frac{(0.1774\ \text{m})}{(83.5 \times 10^{3}\ \text{m/s})}$$
$$= 13.5177\ \mu\text{s}$$

where

$r = 0.1774$m

$U = 83.5 \times 103$ m/s

10.2 ELECTRIC AND MAGNETIC FIELDS COMBINED

When both fields are present in a region at the time, the force on a particle is given by:

$$\mathbf{F} = Q(\mathbf{E} + \mathbf{U} \times \mathbf{B}) \tag{10.7}$$

This force, together with the initial conditions, determines the path of the particle (see Example 10.3) [1].

Example 10.3

A region contains a magnetic flux density $\mathbf{B} = 5.0 \times 10^{-4}\ \mathbf{a_z}$ T and an electric field $\mathbf{E} = 5.0\ \mathbf{a_z}$ V/m. A proton ($Q_p = 1.602 \times 10^{-19}$ C, $m_p = 1.673 \times 10^{-27}$ kg) enters the fields at the origin with an initial velocity $\mathbf{U_o} = 2.5 \times 15^5\ \mathbf{a_x}$ m/s. Describe the proton's motion and give its position after three complete revolutions.

Solution: The initial force on the particle is:

$$\mathbf{F} = Q\left(\mathbf{E} + \mathbf{U}_0 \times \mathbf{B}\right) = Q_p\left(E\mathbf{a}_z - U_0\mathbf{a}_y\right)$$

The z-component (electric component) of the force is constant and produces a constant acceleration in the z-direction. Thus, the equation of motion in the z-direction is:

$$z = \frac{1}{2}at^2 = \frac{1}{2}\left(\frac{Q_pE}{m_p}\right)t^2$$

The other magnetic component, which changes into $-Q_pU\,B\mathbf{a}_r$, produces circular motion that is perpendicular to the z-axis with period equal to:

$$T = \frac{2\pi r}{U} = \frac{2\pi m_p}{Q_pB}$$

The resultant motion is helical as shown in Figure 10.2.

After three revolutions, $x = y = 0$ and

$$z = \frac{1}{2}\left(\frac{Q_pE}{m_p}\right)(3T)^2 = \frac{18\pi^2 Em_p}{Q_pB^2} = 37.0\text{m}$$

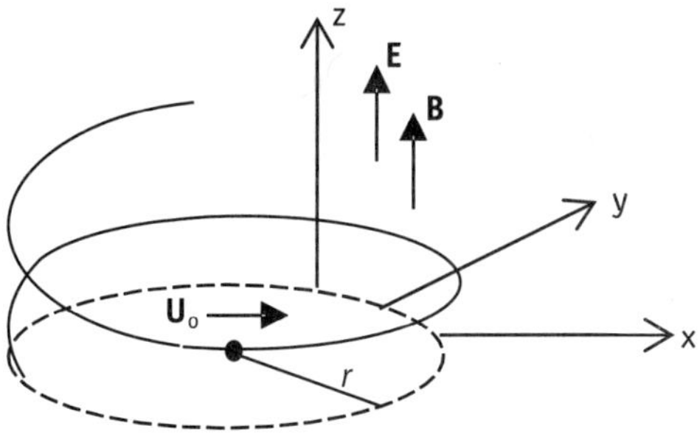

Figure 10.2 Illustration of electric and magnetic fields combined.

10.3 MAGNETIC FORCE ON A CURRENT ELEMENT

A frequently encountered situation is that of a current-carrying conductor in an external magnetic field. Since

$$I = \frac{dQ}{dt} \tag{10.8}$$

the different force equation may be written

$$d\mathbf{F} = dQ(\mathbf{U} \times \mathbf{B}) = (I\,dt)(\mathbf{U} \times \mathbf{B}) = I(d\mathbf{l} \times \mathbf{B}) \tag{10.9}$$

where:

$$d\mathbf{l} = \mathbf{U}dt \tag{10.10}$$

is the elementary length in the direction of the conventional current I. If the conductor is straight and the field is constant along it, the differential force may be integrated to give:

$$F = ILB\sin\theta \tag{10.11}$$

The magnetic force is in fact exerted on the electrons that make up the current I. However, since the electrons are confined to the conductor, the force is effectively transferred to the heavy lattice. This transferred force does the work on the whole conductor. While this fact provides a reasonable introduction to the behavior of current-carrying conductors in electric machines, certain essential considerations have been omitted. No mention was made, nor will be made in Section 10.4, of the current source and the energy that would be required to maintain a constant current *I*. Faraday's law of induction (see Section 12.3) was not applied. In electric machine theory, the result will be modified by these considerations [2].

Conductors in motion in magnetic fields are treated again in Chapter 12.

Example 10.4

Find a force on a straight conductor of length 0.30m carrying a current of 5.0A in the $-\mathbf{a}_z$ direction, where a field is $\mathbf{B} = 3.5 \times 10^{-3}\ (\mathbf{a}_x - \mathbf{a}_y)$ T.

Solution: See the expressions given below:

$$
\begin{aligned}
\mathbf{F} &= I(\mathbf{L} \times \mathbf{B}) = I\left[L\left(-\mathbf{a}_z\right) \times B\left(\mathbf{a}_x - \mathbf{a}_y\right)\right] \\
&= (5.0\ \mathrm{A})\left\{\left[(0.30\ \mathrm{m})\right)\left(-\mathbf{a}_z\right)\right] \times \left[\left(3.5 \times 10^{-3}\ \mathrm{T}\right)\left(\mathbf{a}_x - \mathbf{a}_y\right)\right]\right\} \\
&= 7.42 \times 10^{-3}\left[\frac{\left(-\mathbf{a}_x - \mathbf{a}_y\right)}{\left(2^{1/2}\right)}\right]\ \mathrm{N}
\end{aligned}
$$

where

$I = 5.0\mathrm{A}$

$\mathbf{L} = L(-\mathbf{a}_z) = 0.30\mathrm{m}\ (-\mathbf{a}_z)$

$L = 0.30\mathrm{m}$

$\mathbf{B} = B(\mathbf{a}_x - \mathbf{a}_y) = 3.5 \times 10^{-3}\ \mathrm{T}\ (\mathbf{a}_x - \mathbf{a}_y)$

$B = 3.5 \times 10^{-3}\ \mathrm{T}$

The force having a magnitude of 7.42 mN is at right angles to both the B-field and the current direction as shown in Figure 10.3.

10.4 WORK AND POWER

The magnetic forces on the charged particles and current-carrying conductors examined above result from the field. To counter

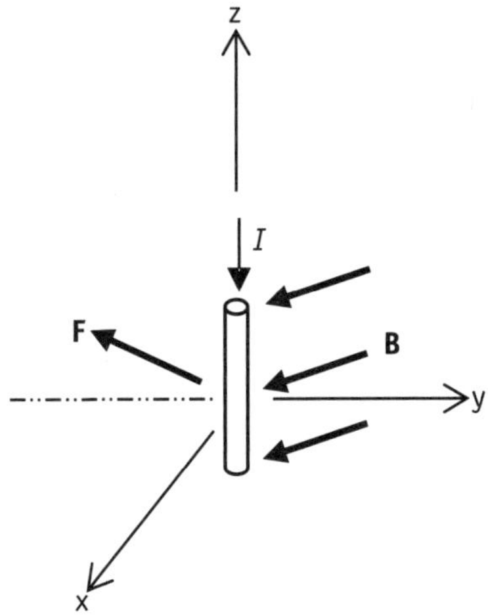

Figure 10.3 Illustration of force on a conductor carrying current.

the forces and establish equilibrium, equal and opposite forces, $\mathbf{F}_a$, would have to be applied. If motion occurs, the work done on the system by the outside agent applying the force is given by the integral:

$$W = \int_{\text{initial } l}^{\text{final } l} \mathbf{F}_a \cdot d\mathbf{l} \tag{10.12}$$

A positive result from the integration indicates that work was done by the agent on the system to move the particles or conductor from the initial location to the final location against the field. Because the magnetic force, and, hence, $\mathbf{F}_a$, is generally nonconservative, the entire path of integration joining the initial and final locations of the conductor must be specified [1].

Example 10.5

Find the work and power required to move the conductor shown in Figure 10.4 one full revolution in the direction shown after 0.02s, if $\mathbf{B} = 2.5 \times 10^{-3}\ \mathbf{a}_r$ T and the current is 45.0A [3].

Solution: See the expressions shown immediately below

$$\mathbf{F} = I(\mathbf{L} \times \mathbf{B}) = I\left[l(-\mathbf{a}_z) \times B(\mathbf{a}_r)\right]$$
$$= -1.13 \times 10^{-2}\mathbf{a}_\phi \text{ N}$$

where

$I = 45.0\text{A}$

$\mathbf{L} = l(-\mathbf{a}_z) = 0.10\text{m}\ (-\mathbf{a}_z)$

$l = 0.10\text{m}$

$\mathbf{B} = B(\mathbf{a}_r) = 2.5 \times 10^{-3}\ \text{T}\ (\mathbf{a}_r)$

$B = 2.5 \times 10^{-3}\ \text{T}$

and so ...

$$\mathbf{F}_a = -1.13 \times 10^{-2}\,\mathbf{a}_\phi\ \text{N}$$

$$W = \int \mathbf{F}_a \cdot d\mathbf{l} = \int_{\phi_{\text{initial}}}^{\phi_{\text{final}}} \mathbf{F}_a \cdot d\mathbf{l}$$

$$= \int_0^{2\pi} \left(-1.13 \times 10^{-2}\right)\mathbf{a}_\phi \cdot \left(r\,d\phi\right)\mathbf{a}_\phi = -2.13 \times 10^{-3}\ \text{J}$$

where

$I = 45.0\text{A}$

$\phi_{\text{final}} = 2\pi \text{ radians} = 2\pi$

$\mathbf{L} = l(-\mathbf{a}_z) = 0.10\text{m}\ (-\mathbf{a}_z)$

$\phi_{\text{initial}} = 0 \text{ radians} = 0$

$l = 0.10\text{m}$

$\mathbf{B} = B(\mathbf{a}_r) = 2.5 \times 10^{-3}\ \text{T}\ (\mathbf{a}_r)$

$B = 2.5 \times 10^{-3}\ \text{T}$

and

$$P = \frac{W}{t} = \frac{-2.13 \times 10^{-3}\ \text{J}}{0.02\ \text{s}} = -0.107\text{W}$$

where

$W = -2.13 \times 10^{-3}\ \text{J}$

$t = 0.02\text{s}$

The negative sign means that work is done by the magnetic field in moving the conductor in the direction shown in Figure 10.4. For motion in the opposite direction, the reversed limits will provide the change of sign, and no attempt to place a sign on $rd\phi\ \mathbf{a}_\phi$ should be made [1].

10.5 TORQUE

The *moment of a force* or *torque* about a specified point is the cross-product of the *lever arm* about that point and the force. The lever arm, **r**, is directed from the point about which the torque is to be obtained to the point of application of the force. In Figure 10.5, the force at P has a torque about the origin given by:

$$\mathbf{T} = \mathbf{r} \times \mathbf{F} \tag{10.13}$$

where **T** has the units Nm (the units Nm/rad is another alternative unit to distinguish torque from energy).

In Figure 10.5, **T** lies along an axis (in the *xy*-plane) through the origin. If P were joined to the origin by a rigid rod freely pivoted at the origin, then the applied force would tend to rotate about that axis. The torque **T** would then be said to be *about the axis* rather than *about point O* [1].

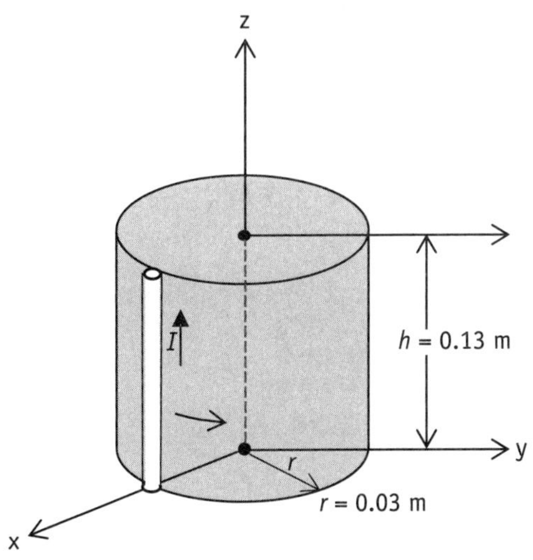

Figure 10.4 Illustration for Example 10.5 to show work and power required.

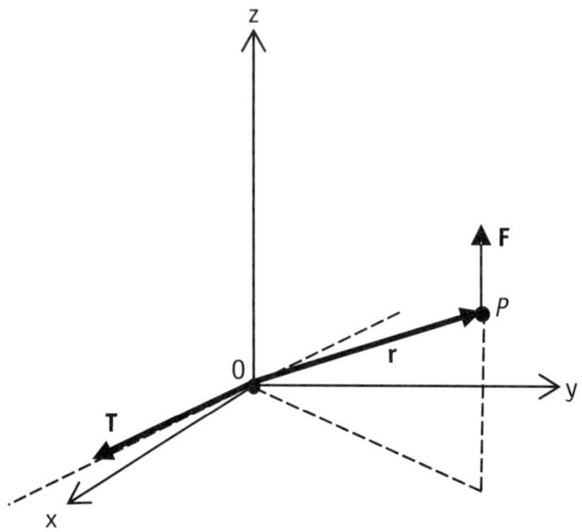

Figure 10.5 Illustration showing force at P having a torque **T** about the origin.

10.6 MAGNETIC MOMENT OF A PLANAR COIL

Consider the single-turn coil in the $z = 0$ plane shown in Figure 10.6 of width, w, in the x-direction and length, l, along y. The field $\mathbf{B}$ is uniform and in the positive x direction. The only forces result from the coil sides, l. For the side on the left,

$$\mathbf{F} = I\left(l\mathbf{a}_y \times B\mathbf{a}_x\right) = -BIl\,\mathbf{a}_x \tag{10.14}$$

and for the side on the right,

$$\mathbf{F} = BIl\,\mathbf{a}_x \tag{10.15}$$

The torque about the y-axis from the left current element requires a lever arm,

$$\mathbf{r} = -\left(\frac{w}{2}\right)\mathbf{a}_x \tag{10.16}$$

The sign will change for the lever arm to the right current element. The torque from both elements is:

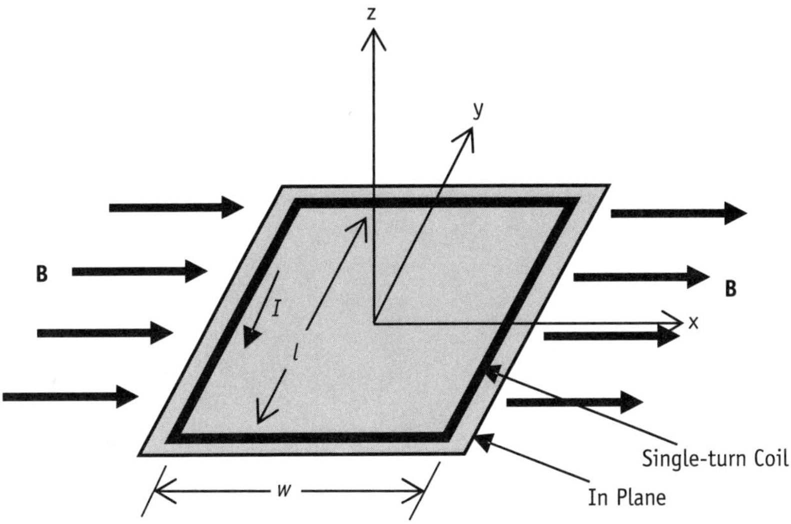

Figure 10.6 Illustration showing single-turn planar coil in the $z = 0$ plane.

$$\begin{aligned}\mathbf{T} &= \left(-\frac{w}{2}\right)\mathbf{a}_x \times (-BI\,l)\mathbf{a}_x + \left(\frac{w}{2}\right)\mathbf{a}_x \times (BI\,l)\mathbf{a}_x \\ &= BI\,l\,w\left(-\mathbf{a}_y\right) = BI\,A\left(-\mathbf{a}_y\right)\end{aligned} \tag{10.17}$$

where A is the area of the coil. It can be shown that this expression for the torque holds for a coil of arbitrary shape (and for any axis parallel to the y-axis).

The *magnetic momentum* **m** of a planar current loop is defined as $I\,A\,\mathbf{a}_n$, where the unit normal $\mathbf{a}_n$ is determined by the right-hand rule. (Note: The right thumb gives the direction of $\mathbf{a}_n$ when the fingers point in the direction of the current.) It is seen that the torque on a planar coil is related to the applied field by:

$$\mathbf{T} = \mathbf{m} \times \mathbf{B} \tag{10.18}$$

This concept of magnetic moment is essential to an understanding of the behavior of orbiting charged particles. For example, a positive charge Q moving in a circular orbit at a velocity, U, or an angular velocity, ω, is equivalent to a current

$$I = \left(\frac{\omega}{2\pi}\right)Q \tag{10.19}$$

and so gives rise to a magnetic moment

$$\mathbf{m} = \left(\frac{\omega}{2\pi}\right)QA\mathbf{a}_{\mathrm{n}} \tag{10.20}$$

as shown in Figure 10.7.

More important to the discussion is the fact that, in the presence of a magnetic field, **B**, there will be a torque **T** = **m** × **B**. This torque tends to turn the current loop until **m** and **B** are in the same direction. In that orientation, the torque will be zero [4].

10.7 PROBLEMS AND SOLUTIONS

Problem 10.1

A conductor 4m long lies along the y-axis with a 10.0A current running in the positive y direction. Find the force on the conductor if the field in the region is **B** = 0.05**x** Tesla.

Problem 10.2

A current strip 2-cm wide carries a current of 15.0A in the x-direction as shown in Figure 10.8.

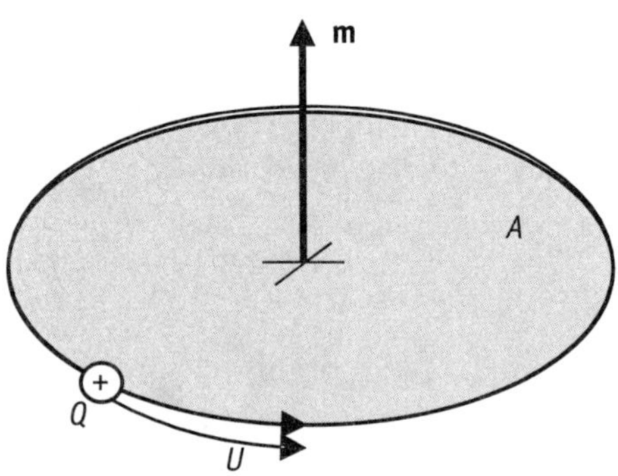

Figure 10.7 Illustration showing positive charge moving about in a circular orbit at a velocity, U, or an angular velocity, ω.

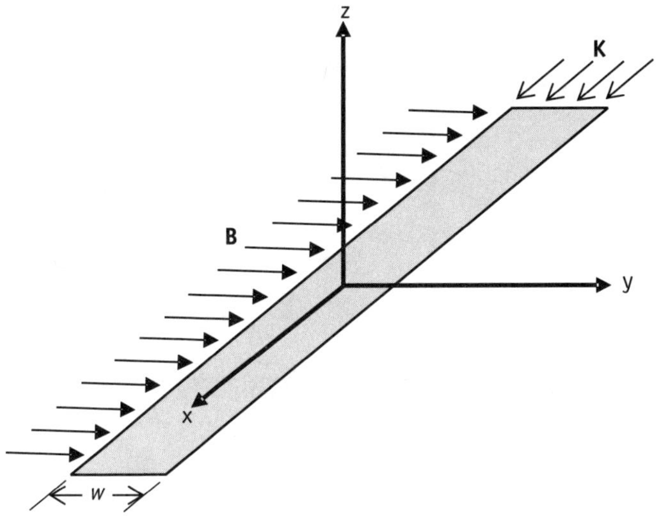

Figure 10.8 Illustration for Problem 10.2.

Find the force on the strip per unit length if the uniform B-field is $\mathbf{B} = 0.20\mathbf{y}$ T.

Problem 10.3

Find the forces per unit length on two long, straight, parallel conductors if each wire carries a current of 10.0A in the same direction. The separation distance between the wires is 0.2m. See Figure 10.9.

Problem 10.4

A conductor carries current, I, parallel to a conductive strip of current density, K_o, and width, w, as shown in Figure 10.10.

Find an expression for the force per unit length on the conductor. What is the result when the width, w, approaches ∞?

Problem 10.5

Find the torque about the y-axis for the two conductors of length, l, separated by a fixed distance, w, in the uniform field, $\mathbf{B}$, shown in Figure 10.11.

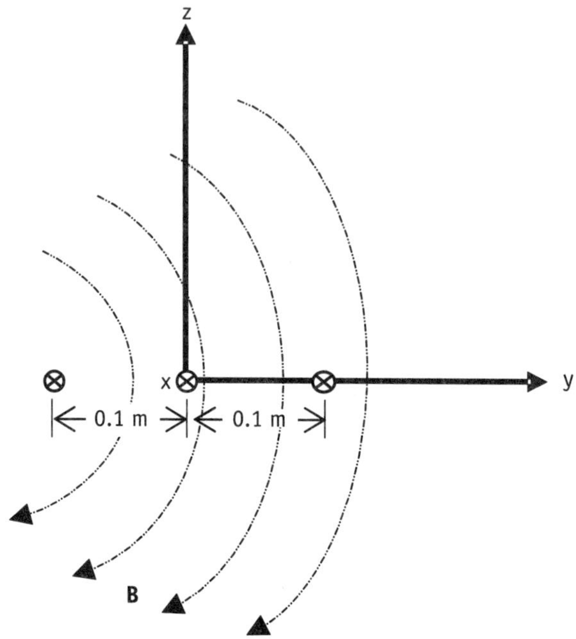

Figure 10.9 Illustration for Problem 10.3.

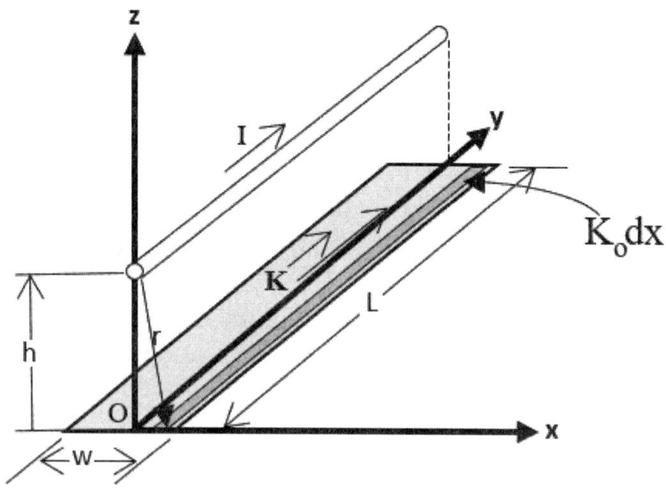

Figure 10.10 Illustration for Problem 10.4.

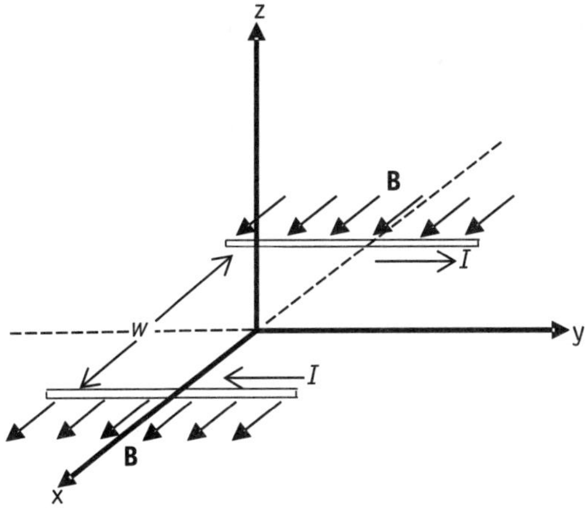

Figure 10.11 Illustration for Problem 10.5.

Problem 10.6

The rectangular coil in Figure 10.12 is in the field quantified as:

$$\mathbf{B} = 0.05\ (1/2^{1/2})\ (\mathbf{x} + \mathbf{y})\ \text{Tesla}$$

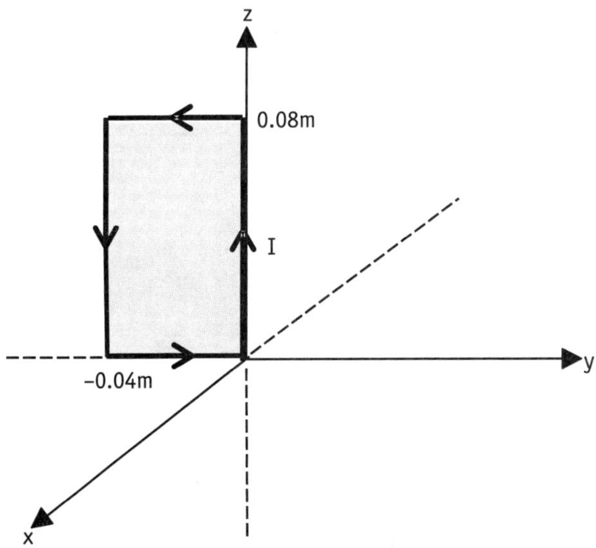

Figure 10.12 Illustration for Problem 10.6.

Find the torque about the z-axis when the coil is in the position, as shown in the figure, and carries a current of 5.0A.

Problem 10.7

Find the maximum torque on an 85-turn, rectangular coil measuring 0.2m by 0.3m and carrying a current of 2.0A in a field quantified to be B = 6.5 Tesla.

Problem 10.8

A conductor 4m long carrying a current of 10A in the $+y$ direction lies along the y-axis between the $y = \pm 2$m. If the field $\mathbf{B} = 0.05\mathbf{x}$ Tesla, find the work done moving the conductor parallel to itself at a constant speed to the position $x = z = 2.0$m. See Figure 10.13.

Problem 10.9

Find the work and power required to move the conductor shown in Figure 10.14 below one full turn in the positive direction at rotational frequency of N revolutions per minute if $\mathbf{B} = B_o\,\mathbf{r}$ Tesla (note that B_o is a positive constant).

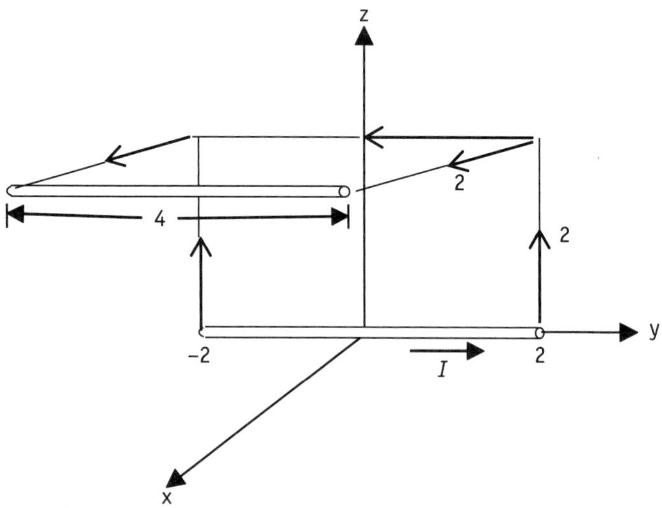

Figure 10.13 Illustration for Problem 10.8.

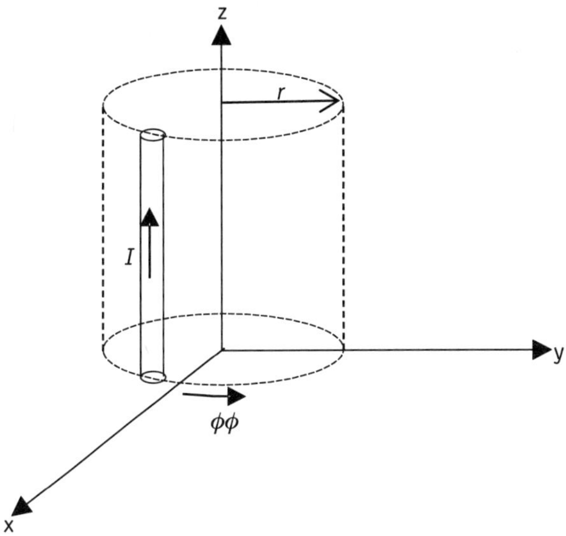

Figure 10.14 Illustration for Problem 10.9.

References

[1] Shadowitz, A., *The Electromagnetic Field*, New York: Dover Publications, 1975.

[2] Russer, P., *Electromagnetics, Microwave Circuit and Antenna Design for Communications Engineering*, Norwood, MA: Artech House, 2006.

[3] Warnick, K. F., and P. Russer, *Problem Solving in Electromagnetics, Microwave Circuit and Antenna Design for Communications Engineering*, Norwood, MA: Artech House, 2006.

[4] Cheng, D. K., *Fundamentals of Engineering Electromagnetics*, Reading, MA: Addison-Wesley, 1993.

11

INDUCTANCE AND MAGNETIC CIRCUITS

11.1 VOLTAGE OF SELF-INDUCTION

A voltage appears at the terminals of an N-turn coil in Figure 11.1, where the flux ϕ common to the turns is changing with time. This induced voltage is given by Faraday's law:

$$V = -N\left(\frac{d\phi}{dt}\right) \tag{11.1}$$

See Chapter 12 for a discussion of the polarity. Defining the *self-inductance* of the coil by:

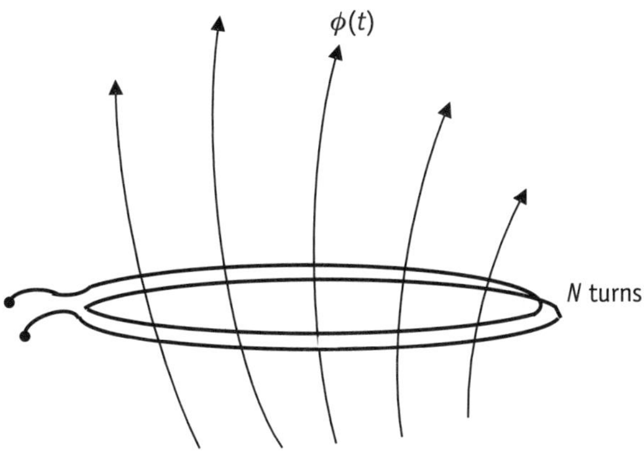

Figure 11.1 N-turn coil with flux ϕ.

$$L = N\left(\frac{d\phi}{di}\right) \tag{11.2}$$

Faraday's law may be written in the form:

$$V = -L\left(\frac{di}{dt}\right) \tag{11.3}$$

The units on L are henries where:

$$1 \text{ H} = 1 \text{ Wb/A} \tag{11.4}$$

In any calculations of self-inductance, account must be taken of the ferromagnetic material presence [1]. The flux, and, hence, the self-inductance, when free space is assumed will differ greatly from that which results when a portion of the region contains ferromagnetic materials.

11.2 INDUCTORS AND INDUCTANCE

An inductor (also called an inductance) is formed by two conductors separated by free space, where the conductor arrangement is such that magnetic flux from one links with the other. For static (or, at most, low-frequency) current I in the conductors, let the total flux linking the conductors be:

$$\lambda = -\left| \begin{array}{ll} N\Phi & \text{for coils} \\ \Phi & \text{for other arrangements} \end{array} \right. \tag{11.5}$$

Then, the inductance of the inductor is defined to be:

$$L = \frac{\lambda}{I} \tag{11.6}$$

It should be noted that L will always be the product of μ_0 and a geometrical factor having the dimensions of length. Compare with expressions for resistance, R (Chapter 6) and capacitance, C (Chapter 7) [1].

The expressions for the stored energy in the magnetic field from circuit analysis and field theory provide another defining expression for inductance, L. The equations [2],

$$W = \frac{1}{2} L I^2 \tag{11.7}$$

and [3],

$$W = \frac{1}{2} \int_V (\mathbf{B} \cdot \mathbf{H})\, dv \tag{11.8}$$

give:

$$L = \int_V (\mathbf{B} \cdot \mathbf{H})\, dv / I^2 \tag{11.9}$$

11.3 STANDARD FORMS

In addition to coaxial conductors, several of the more common conductor configurations for which the inductance is required are shown in Figures 11.2 through 11.8 [2].

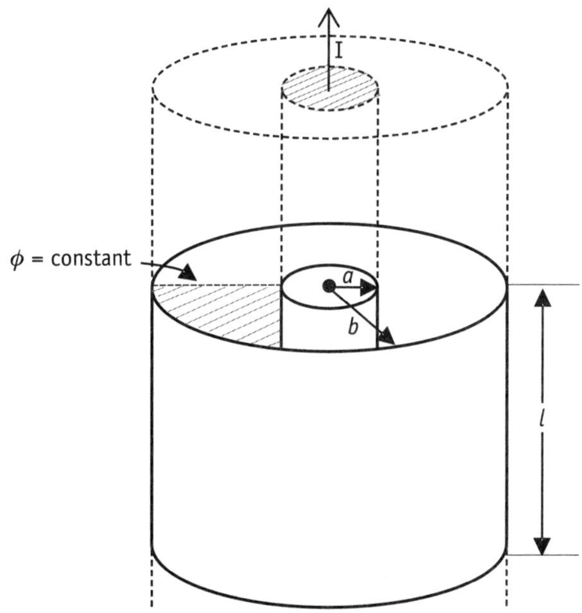

Figure 11.2 Illustration of coaxial conductors.

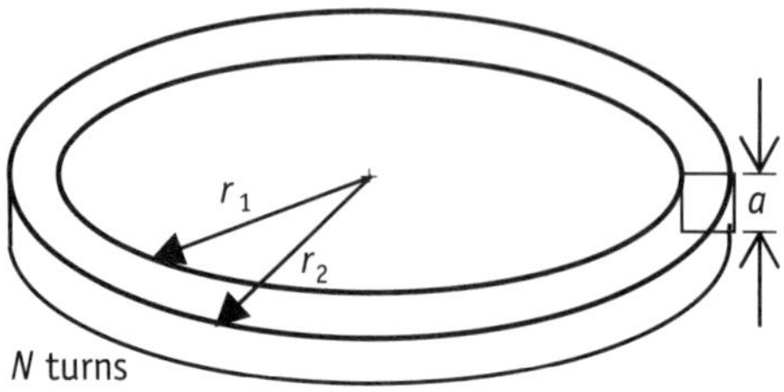

Figure 11.3 Illustration of toroid, square cross-section.

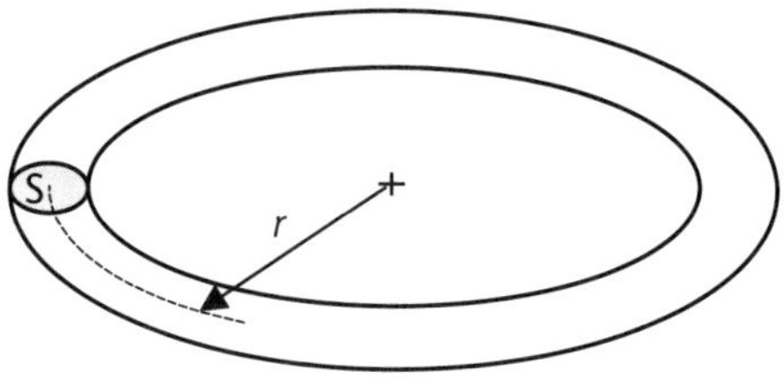

Figure 11.4 Illustration of toroid, general cross section *S*.

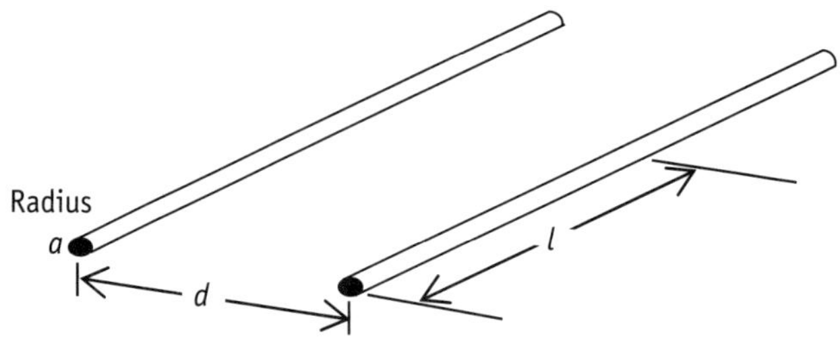

Figure 11.5 Illustration of parallel conductors of radius *a*.

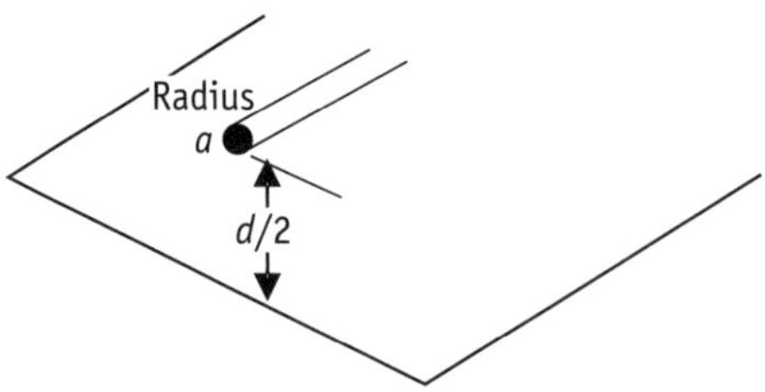

Figure 11.6 Illustration of cylindrical conductor parallel to a ground.

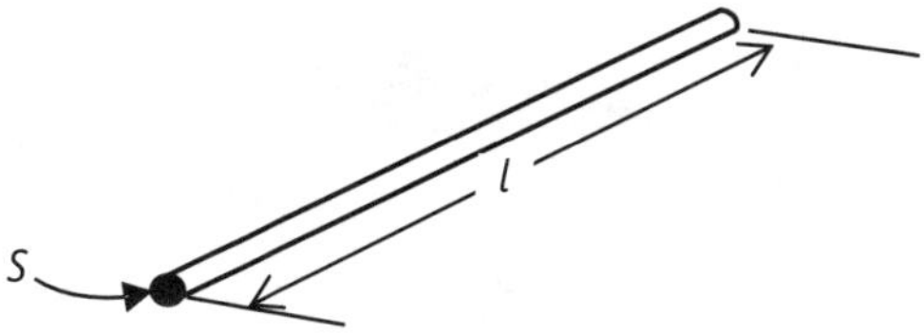

Figure 11.7 Long solenoid of small cross-sectional area, *S*.

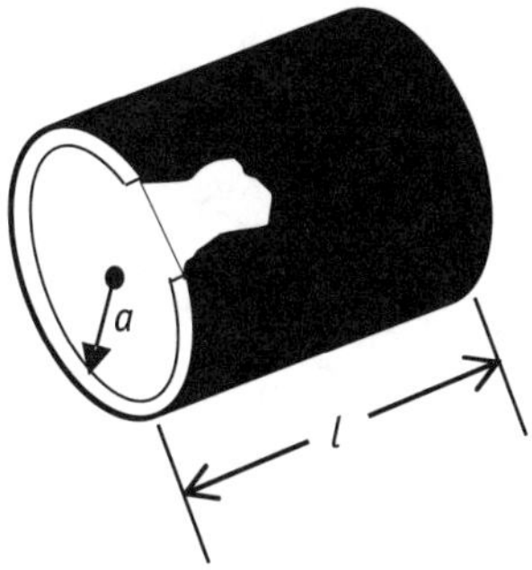

Figure 11.8 Single layer, air-core coil.

11.4 INTERNAL INDUCTANCE

Magnetic flux occurs within a conductor cross section as well as external to the conductor. This internal flux gives rise to an *internal inductance* which is often small compared to the external inductance and frequency ignored. In Figure 11.9 and Figure 11.10(a, b) a conductor of circular cross section is shown, with a current I assumed to be uniformly distributed over the area. (Note: this assumption is only valid at low frequencies seeing how *skin effect* at higher frequencies forces the current to be concentrated at the outer surface.) Within the conductor of radius, a, Ampere's law gives

$$\mathbf{H} = \left(\frac{Ir}{2\pi a^2}\right)\mathbf{a}_\phi \tag{11.10}$$

and,

$$\mathbf{B} = \mu_0\left(\frac{Ir}{2\pi a^2}\right)\mathbf{a}_\phi \tag{11.11}$$

Figure 11.9 Multilayer, air-core coil.

The piece of conductor in Figure 11.10(a) must be imagined as a short section of an infinite torus as suggested in Figure 11.10(b). The current filaments become circles of infinite radius. The lines of flux $d\Phi$ through the strip $l\ dr$ encircle only those filaments whose distance from the conductor axis is smaller than r. Thus, an open surface bounded by one of those filaments is cut once (or an odd number of times) by the lines of $d\Phi$; whereas for a filament such as *1* or *2*, the surface is cut zero times (or an even number of times). It follows that $d\Phi$ links only with the fraction $\pi r^2/\pi a^2$ of the total current so that the total flux linkage is given by the weighted sum,

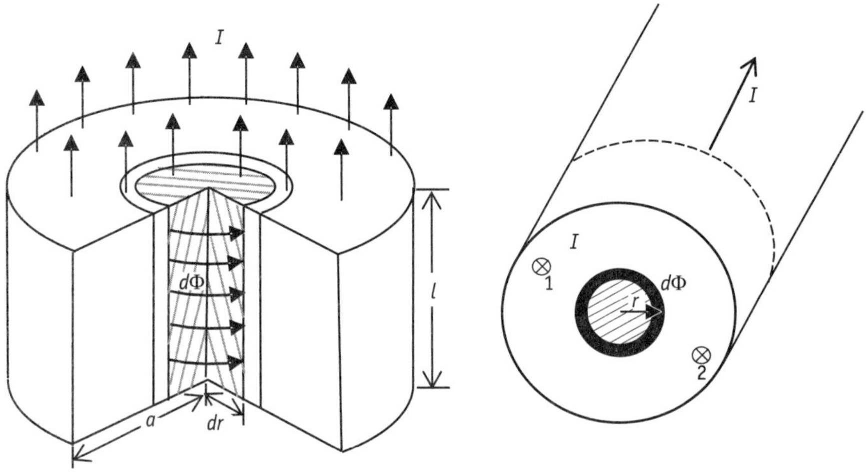

Figure 11.10 Conductor of circular coaxial cross section.

$$\lambda = \int\left(\frac{\pi r^2}{\pi a^2}\right)d\Phi = \int_0^a\left(\frac{\pi r^2}{\pi a^2}\right)\left(\frac{\mu_0 Ir}{2\pi a^2}\right)l\,dr = \left(\frac{\mu_0 Il}{8\pi}\right) \tag{11.12}$$

and

$$\frac{L}{l} = \frac{(\lambda I)}{l} = \frac{\mu_0}{8\pi} = \frac{1}{2}\times 10^{-7}\text{ H/m} \tag{11.13}$$

The result is independent of the conductor radius. The total inductance is the sum of the external and internal inductances. If the external inductance is the of the order $1/2 \times 10^{-7}$ H/m, the internal inductance should not be ignored [3].

11.5 MAGNETIC CIRCUITS

In Chapter 9, magnetic field intensity **H**, flux **Φ**, and magnetic flux density **B** were examined where the medium was free space. For example, when Ampere's law is applied to the closed path C through the long, air-core shown in Figure 11.11.

The result is:

$$\oint \mathbf{H}\cdot d\mathbf{l} = NI \tag{11.14}$$

But since the flux lines are widely spread outside of the coil, B is small there. The flux is effectively restricted to the inside of the coil where:

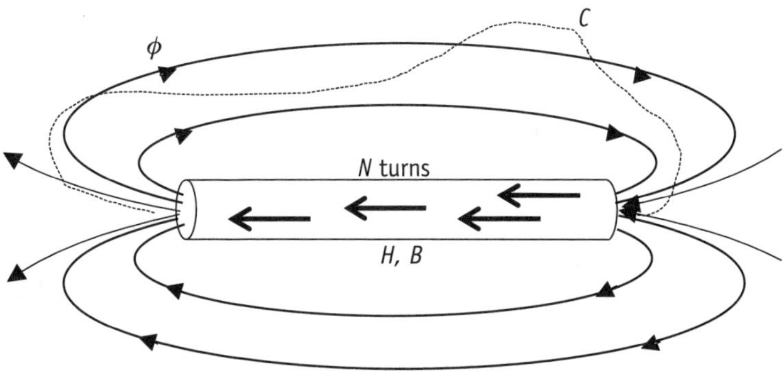

Figure 11.11 Ampere's law applied to the closed path, C.

$$H \cong \frac{NI}{l} \tag{11.15}$$

Ferromagnetic materials have relative permeabilities, μ_r, in the order of thousands.

Consequently, the flux density:

$$B = \mu_0 \mu_r H \tag{11.16}$$

is for a given H much greater than would result in free space. In Figure 11.12, the coil is not distributed over the entire iron core. Even so, the NI of the coil causes a flux Φ which follows the core. It might be said that the flux prefers the core to the surrounding space by the ratio of several thousand to one.

This is so different from the free-space magnetics of Chapter 9 that an entire subject area, known as iron core magnetics or magnetic circuits, was developed. This brief introduction to the subject assumes that all of the flux is within the core. It is also assumed that the flux is uniformly distributed over the core's cross section. Core lengths required for calculation of NI drops are mean lengths [4].

11.6 NONLINEARITY OF THE B-H CURVE

A sample of ferromagnetic material could be tested by applying increasing values of H and measuring the corresponding values of flux density B. *Magnetization curves*, or simply, *B-H curves*, for some common ferromagnetic materials are given in Figures 11.13 and 11.14.

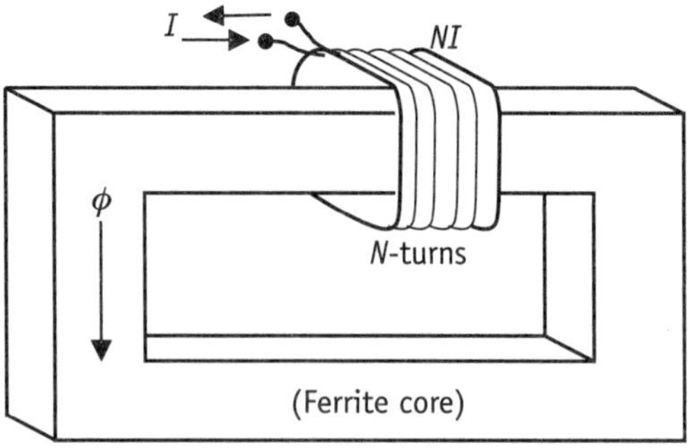

Figure 11.12 Electric conductors wrapped around a ferromagnetic core.

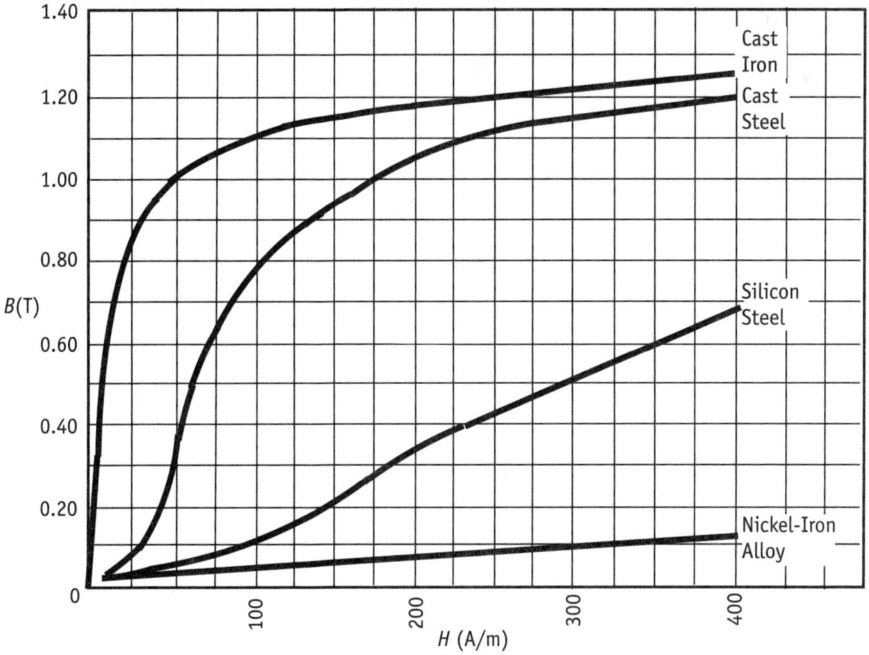

Figure 11.13 Magnetization (B-H) curves for four materials. $H < 400\ A/M$.

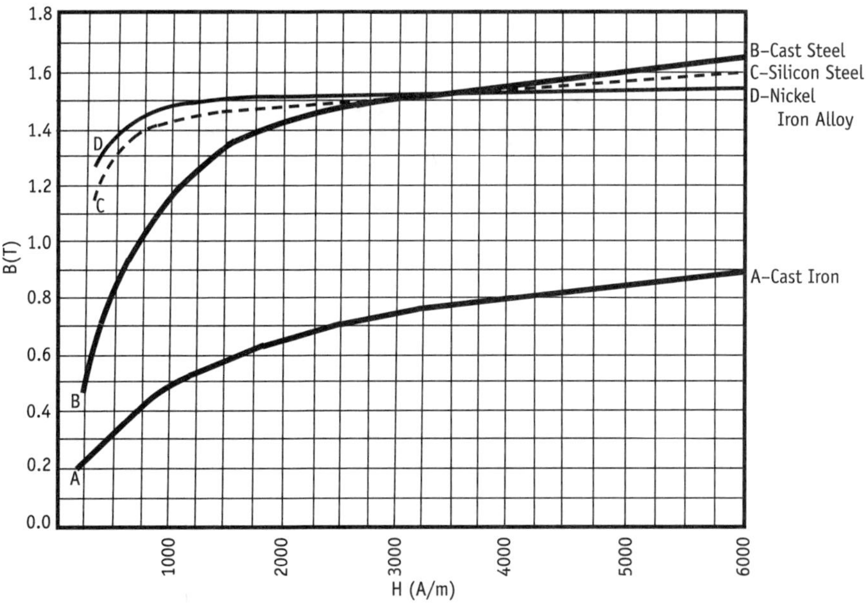

Figure 11.14 Magnetization (B-H) curves for four materials. $H > 400\ A/M$.

The relative permeability can be computed from the *B-H* curve by use of both the curve and the equation

$$\mu_r = \frac{B}{(\mu_0 H)} \tag{11.17}$$

Figure 11.15 shows the extreme nonlinearity of μ_r versus H for silicon steel. This nonlinearity requires that problems be solved graphically [1].

11.7 AMPERE'S LAW FOR MAGNETIC CIRCUITS

A coil of *N* turns and current *I* around a ferromagnetic core produces a *magnetomotive force* (mmf) given by *NI*. The symbol *F* is sometimes

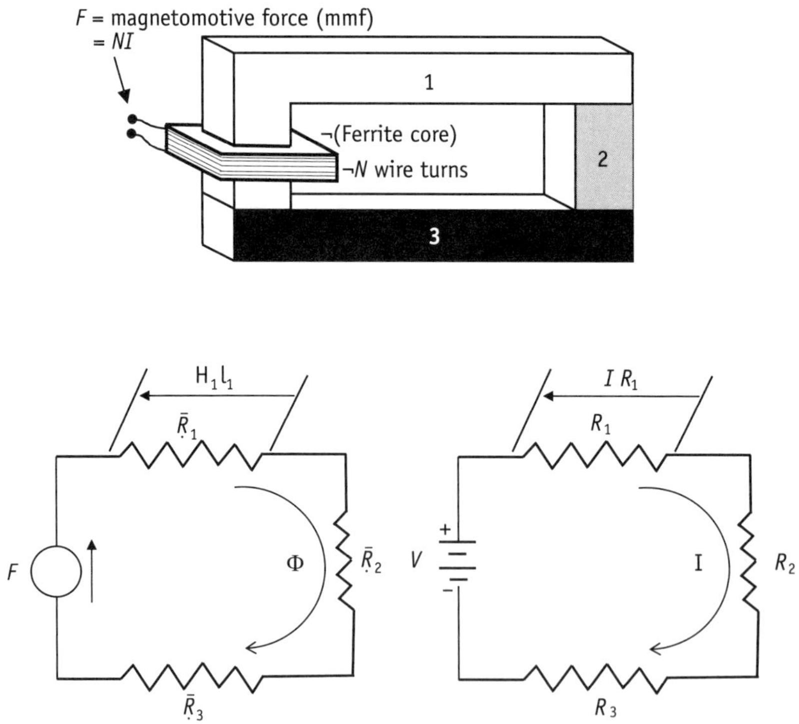

Figure 11.15 Ampere's law for magnetic circuits.

used for this mmf. The units are amperes (A) Ampere's law applied around the path in the center of the core shown in Figure 11.15(a), gives:

$$\begin{aligned} F = NI &= \oint \mathbf{H} \cdot d\mathbf{l} = \int_1 \mathbf{H} \cdot d\mathbf{l} = \int_2 \mathbf{H} \cdot d\mathbf{l} = \int_3 \mathbf{H} \cdot d\mathbf{l} \\ &= H_1 l_1 + H_2 l_2 + H_3 l_3 \end{aligned} \tag{11.18}$$

Comparison with Kirchhoff's law around a single closed loop with three resistors and an emf V,

$$V = V_1 + V_2 + V_3 \tag{11.19}$$

suggests that F can be viewed as an NI rise and the Hl terms considered NI drops in analogy to the voltage rise V and voltage drops V_1, V_2, and V_3. The analogy is developed in Figures 11.15(b) and 11.15(c). Flux Φ in Figure 11.15(b) is analogous to current I (see Figure 11.15(c)).

Reluctance R is analogous to electrical resistance R. An expression for reluctance can be developed as follows:

$$NI \text{ drop} = Hl = BA\left(\frac{l}{\mu A}\right) = \Phi \bar{R} \tag{11.20}$$

Hence,

$$\bar{\mathbf{R}} = \left[\frac{l}{(\mu A)}\right] \left(H^{-1}\right) \tag{11.21}$$

If the reluctances are known, then the equation

$$F = NI = \Phi\left(\bar{R}_1 + \bar{R}_2 + \bar{R}_3\right) \tag{11.22}$$

can be written for the magnetic circuit of Figure 11.15(b). However, μ_r must be known for each material before reluctance can be calculated, and only after B or H is known will the value of μ_r be known. This is in contrast to the relation

$$R = \frac{l}{\sigma A} \tag{11.23}$$

See Section 6.7 in which the conductivity, σ, was independent of the current.

11.8 CORES WITH AIR GAPS

Magnetic circuits with small air gaps are very common. The gaps are generally kept as small as possible since the *NI* drop of the air gap is often much greater than the drop in the core. The flux fringes outward at the gap so that the area at the gap exceeds the area of the adjacent core. Provided that the gap length, l_a, is less than 1/10 the smaller dimension of the core, an *apparent area*, S_a, of the air gap can be calculated. For a rectangular core of dimensions a and b,

$$S_a = (a + l_a)(b + l_a) \tag{11.24}$$

If the total flux in the air gap is known, H_a and $H_a l_a$ can be computed directly.

$$H_a = \frac{1}{\mu_0}\left(\frac{\Phi}{S_a}\right) \tag{11.25}$$

$$H_a l_a = l_a\left(\frac{1}{\mu_0}\right)\left(\frac{\Phi}{S_a}\right) = \left(\frac{l_a \Phi}{\mu_0 S_a}\right) \tag{11.26}$$

For a uniform iron core of length, l_i, with a single air gap, Ampere's law reads:

$$NI = H_i l_i + H_a l_a = H_i l_i + \frac{l_a \Phi}{\mu_0 S_a} \tag{11.27}$$

If the flux Φ is known, it is not difficult to compute the *NI* drop across the air gap, obtain B_i, take H_i from the appropriate *B-H* curve and compute the *NI* drop in the core, $H_i l_1$. The sum is the *NI* required to establish the flux Φ. However, with *NI* given, it is a matter of trial and error to obtain B_i and Φ.

11.9 MULTIPLE COILS

Two or more coils on a core could be wound such that their mmfs either aid one another or oppose. Consequently, a method of indicating polarity is given in Figure 11.16.

An assumed direction for the resulting flux Φ could be incorrect just as an assumed current in a dc circuit with two or more voltage sources may be incorrect. A negative result simply means that the flux is in the opposite direction [3].

11.10 PARALLEL MAGNETIC CIRCUITS

The method of solving a parallel magnetic circuit is suggested by the two-loop equivalent circuit shown in Figure 11.17(a, b).

The leg on the left contains an *NI* rise and an *NI* drop. The *NI* drop between the junctions a and b can be written for each leg as follows:

$$F - H_1 l_1 = H_2 l_2 = H_3 l_3 \tag{11.28}$$

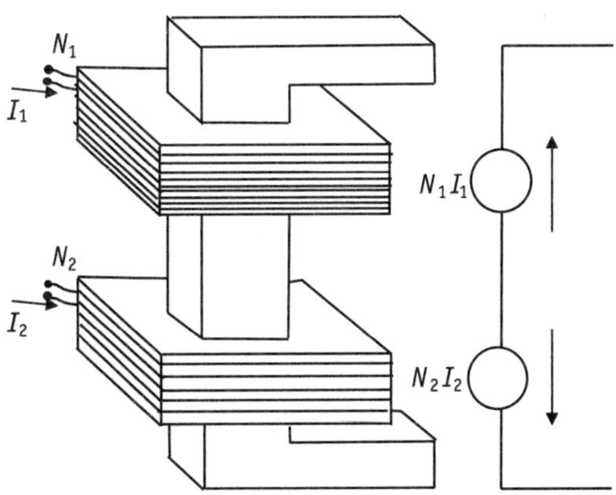

Figure 11.16 Illustration of multiple coils.

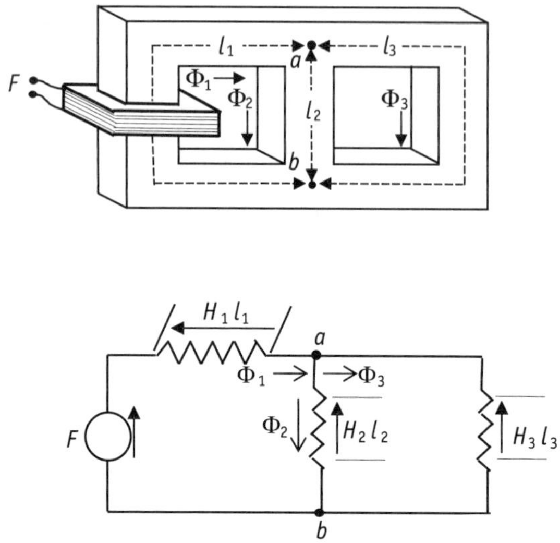

Figure 11.17 Illustration of parallel magnetic circuits.

and the fluxes satisfy

$$\Phi_1 = \Phi_2 + \Phi_3 \tag{11.29}$$

Different materials for the core parts will necessitate working with several *B-H* curves. An air gap in one of the legs would lead to $H_i l_i + H_a l_a$ for the mmf between the junctions for that leg.

The equivalent magnetic circuit should be drawn for parallel magnetic circuit problems. It is good practice to mark the material types, cross-sectional areas, and mean length directly on the diagram. In more complex problems, a scheme like Table 11.1 can be helpful. The data are inserted directly into the table. The remaining

Table 11.1
Data Collection Matrix for Magnetic Circuit Analysis

Part	Material	Area	*l*	Φ	*B*	*H*	*H l*
#1							
#2							
#3							

quantities are then calculated or taken from the appropriate *B-H* curve [4].

11.11 PROBLEMS AND SOLUTIONS

Problem 11.1

Find the inductance per unit length of the coaxial transmission line given in Figure 11.18.

Problem 11.2

Find the inductance per unit length of the parallel cylindrical conductors (see Figure 11.19) where $l = 25$ ft and radius $a = 0.803$ in.

Problem 11.3

A circular conductor with the same radius, a, as in the previous problem, is 12.5 feet from an infinite conducting plane.

Find the inductance (see Figure 11.20).

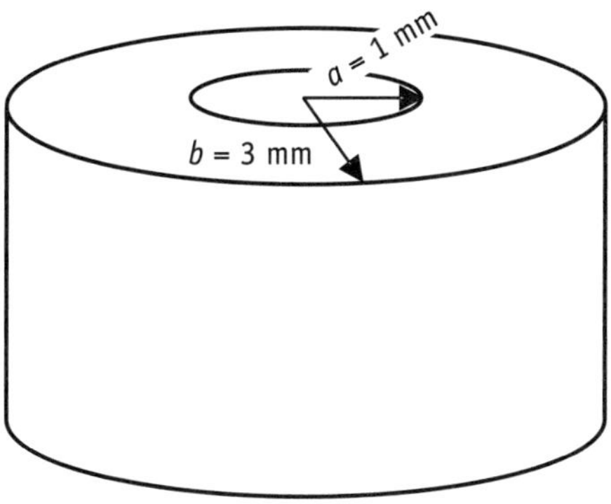

Figure 11.18 Illustration for Problem 11.1.

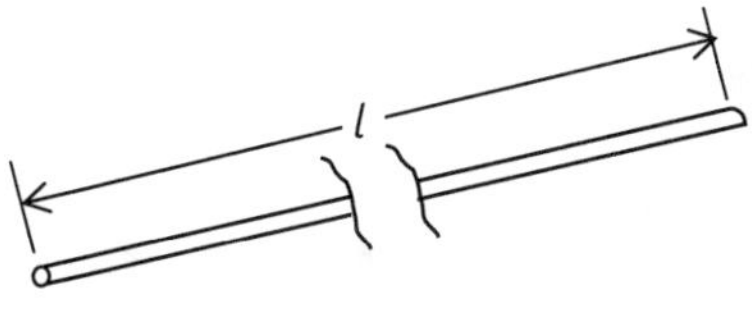

Figure 11.19 Illustration for Problem 11.2.

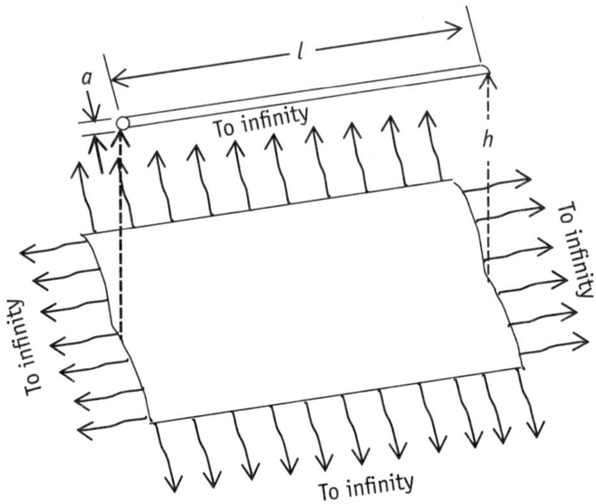

Figure 11.20 Illustration for Problem 11.3.

Problem 11.4

An air-core solenoid of 300 turns and length 0.5m has a single layer of conductor at a radius 0.02m. Find the inductance.

Problem 11.5

Find the inductance of the coil shown in Figure 11.21 where $N = 300$, $r_1 = 9$ mm, $r_2 = 25$ mm, and $l = 20$ mm.

Problem 11.6

Assume that the air-core toroid shown in Figure 11.22 has a circular cross section of radius 4 mm.

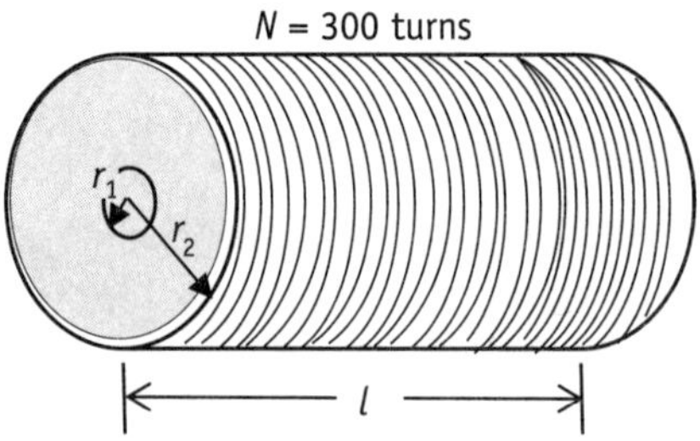

Figure 11.21 Illustration for Problem 11.5.

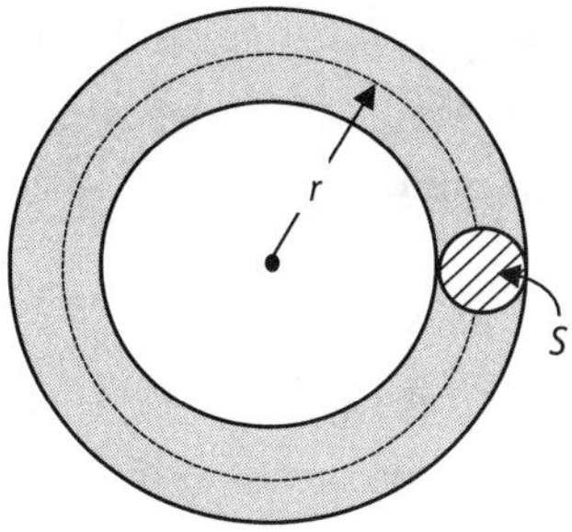

Figure 11.22 Illustration for Problem 11.6.

Find the inductance if there are 2,500 turns and the mean radius is $r = 20$ mm.

Problem 11.7

Assume the air-core toroids of Figure 11.23 have 700 turns, an inner radius of 1 cm, an outer radius of 2 cm, and a height = 1.5 cm. Find L using the appropriate equations.

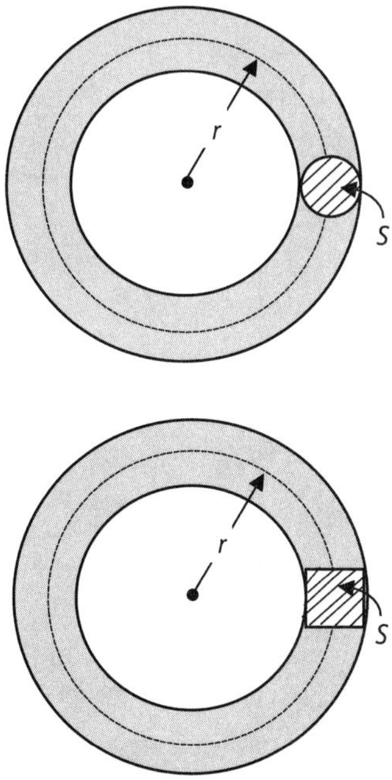

Figure 11.23 Illustration for Problem 11.7.

(a) The formula for square cross-section toroids;

(b) The approximate formula for general toroid which assumes a uniform H at a mean radius.

Problem 11.8

Use the energy integral to find the internal inductance per unit length of a length of a cylindrical conductor of radius a (see Figure 11.24).

Problem 11.9

The cast iron core shown in Figure 11.25 has an inner radius of 7 cm and an outer radius of 9 cm. Find the flux, Φ, if the coil mmf (F) is 500A.

Problem 11.10

The magnetic circuit shown in Figure 11.26 is cast iron with a mean length, l_1, of 0.44m and square cross section of 0.02m × 0.02m. The air gap length, l_a, 2 mm and the coil contains 400 turns.

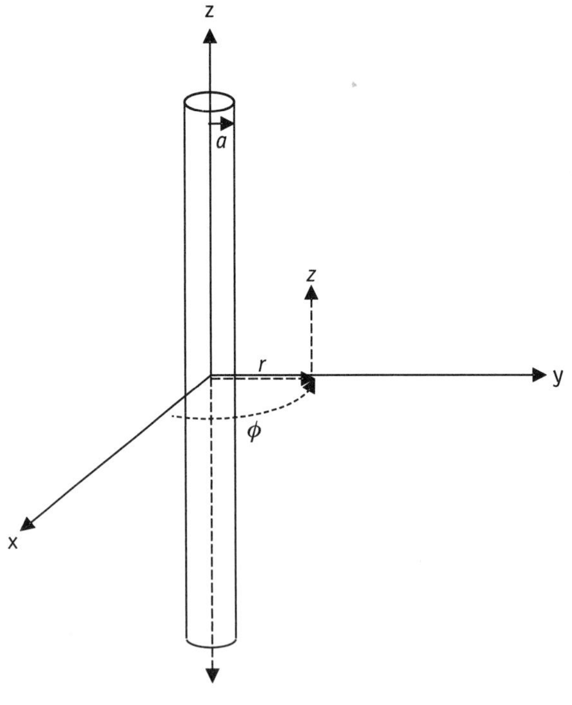

Figure 11.24 Illustration for Problem 11.8.

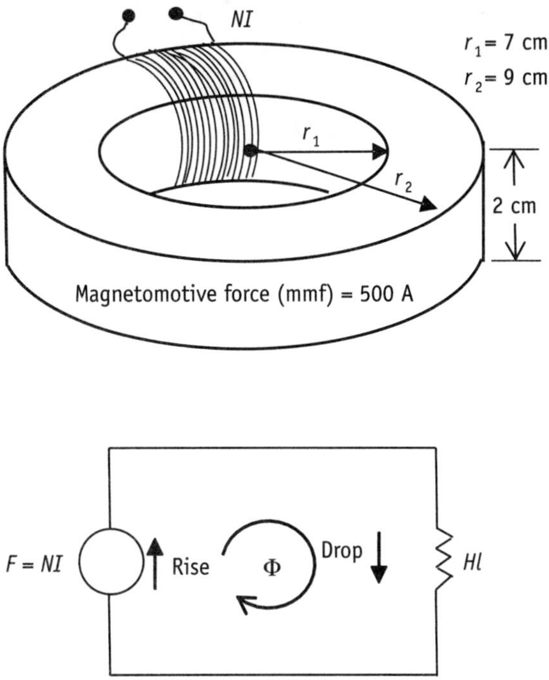

Figure 11.25 Illustration for Problem 11.9.

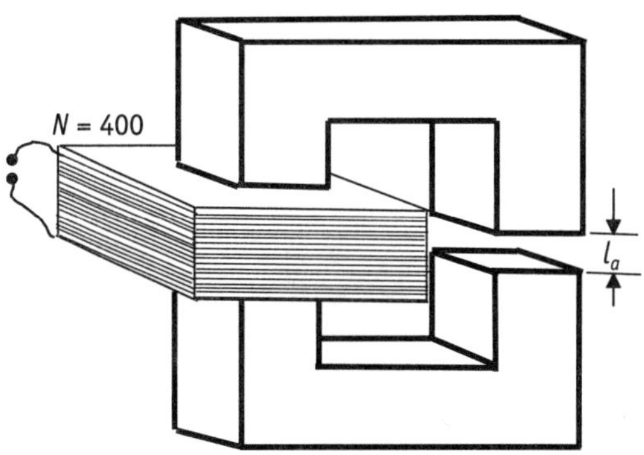

Figure 11.26 Illustration for Problem 11.10.

Find the current I required to establish an air gap flux of 0.141 mWb.

Problem 11.11

Determine the reluctance of an air gap in a dc machine in which the apparent area is:

$$S_a = 4.26 \times 10^{-2} \text{ m}^2$$

and the gap length is $l_a = 5.6$ mm.

References

[1] Shadowitz, A., *The Electromagnetic Field*, New York: Dover Publications, 1975.

[2] Russer, P., *Electromagnetics, Microwave Circuit and Antenna Design for Communications Engineering*, Norwood, MA: Artech House, 2006.

[3] Warnick, K. F., and P. Russer, *Problem Solving in Electromagnetics, Microwave Circuit and Antenna Design for Communications Engineering*, Norwood, MA: Artech House, 2006.

[4] Cheng, D. K., *Fundamentals of Engineering Electromagnetics*, Reading, MA: Addison-Wesley, 1993.

12

DISPLACEMENT CURRENT AND INDUCED EMF

12.1 DISPLACEMENT CURRENT

For static fields, the curl of **H** is equal to **J**, the current density at the point. This current density, which is due to the motion of actual charges (i.e., electrons, protons, and ions), will be called the *conduction current* density and will be denoted $\mathbf{J}_c$. It is clear that, if

$$\nabla \times \mathbf{H} = \mathbf{J}_c \tag{12.1}$$

is also held for time varying fields, then

$$\nabla \cdot \mathbf{J}_c = \nabla \cdot (\nabla \times \mathbf{H}) = 0 \tag{12.2}$$

(See Section 9.4.)

This would be incompatible with the continuity equation,

$$\nabla \cdot \mathbf{J}_c = -\frac{\partial \rho}{\partial t} \tag{12.3}$$

obtained in Section 6.9. James Clerk Maxwell postulated that [1]:

$$\nabla \times \mathbf{H} = \mathbf{J}_c + \mathbf{J}_D \tag{12.4}$$

where the *displacement current* density, $\mathbf{J}_D$, is defined by

$$\mathbf{J}_D = \frac{\partial \mathbf{D}}{\partial t} \tag{12.5}$$

With the displacement current added to the conduction current, the continuity equation is saved. Indeed,

$$\nabla \cdot \mathbf{J}_c = \nabla \cdot (\nabla \times \mathbf{H}) - \nabla \cdot \mathbf{J}_D = 0 - \nabla \cdot \frac{\partial \mathbf{D}}{\partial t} = -\frac{\partial}{\partial t}(\nabla \cdot \mathbf{D}) = -\frac{\partial \rho}{\partial t} \tag{12.6}$$

where $\nabla \cdot \mathbf{D} = \rho$ (Section 4.3) has been utilized.

The displacement current (in amperes) through a specified open surface is obtained by integration and in the same way as is the conduction current. Thus,

$$i_c = \int_S \mathbf{J}_c \cdot d\mathbf{S} \tag{12.7}$$

$$i_D = \int_S \mathbf{J}_D \cdot d\mathbf{S} = \int_S \frac{\partial \mathbf{D}}{\partial t} \cdot d\mathbf{S} \tag{12.8}$$

The expression for i_D may be interpreted in terms of the motion of charges. If charge Q is moving with velocity **U**, the electric field surrounding Q also advances with velocity **U**.

Therefore, even though Q may not be physically crossing a surface S (which would constitute a conduction current), the lines of **D** will be changing across S, resulting in a displacement current [1].

Example Problem

A time-varying voltage applied to a parallel plate capacitor (see Figure 12.1) [2] results in a time-varying current, i_c, in the connecting leads. Two open surfaces with a common contour C are shown [1]. Because $\nabla \cdot (\nabla \times \mathbf{H}) = 0$, the divergence theorem (Section 4.5) gives

$$\int_{S_1} (\nabla \times \mathbf{H}) \cdot d\mathbf{S} = \int_{S_z} (\nabla \times \mathbf{H}) \cdot d\mathbf{S}$$

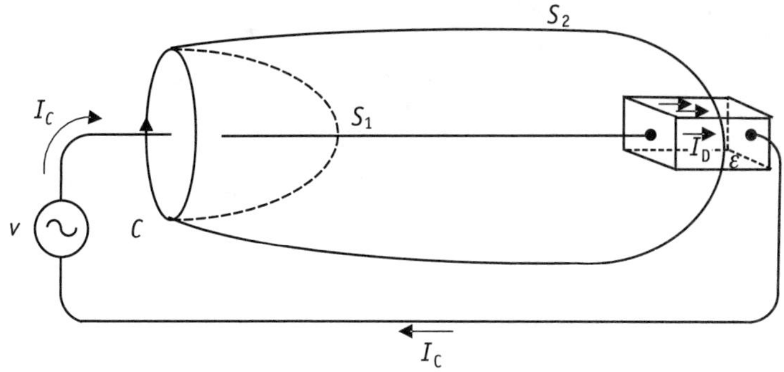

Figure 12.1 Configuration with two open surfaces having a common contour.

or

$$\int_{S_1}\left[\mathbf{J}_c + \left(\frac{\partial \mathbf{D}}{\partial t}\right)\right]\cdot d\mathbf{S} = \int_{S_2}\left[\mathbf{J}_c + \left(\frac{\partial \mathbf{D}}{\partial t}\right)\right]\cdot d\mathbf{S}$$

In the absence of fringing, **D** will be changing only within the capacitor. Moreover, there can be no moving charges ($J_c = 0$) in the dielectric. Therefore,

$$\int_{S_1}\mathbf{J}_c \cdot d\mathbf{S} = \int_{S_2}\left(\frac{\partial \mathbf{D}}{\partial t}\right)\cdot d\mathbf{S}$$

where S_3 is that part of S_2 which lies in the dielectric. The integral on the left is simply i_c, the conduction current, made up of the moving charges in the leads. The integral on the right is the displacement current, i_D, in the dielectric.

The equality of i_c and i_D for this case is demonstrated in Problem 12.1.

12.2 RATIO OF J_C TO J_D

Some materials are neither good conductors nor perfect dielectrics such that both conduction current and displacement current exist. A model for the poor conductor or lossy dielectric is shown in Figure 12.2.

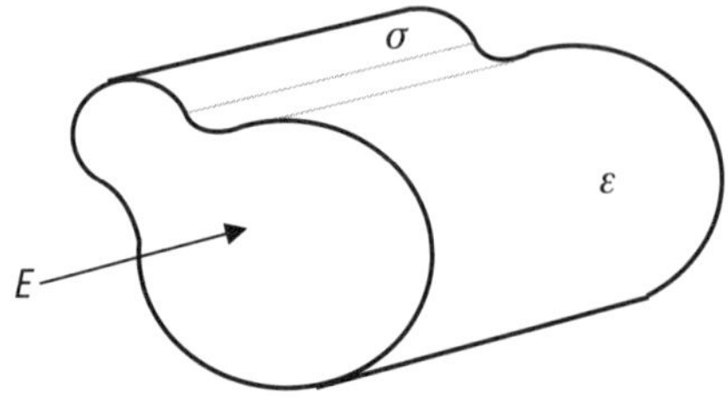

Figure 12.2 Model for the poor conductor or lossy dielectric.

Assuming the time-dependent $e^{j\omega t}$ for **E**, the total current density is:

$$\mathbf{J}_t = \mathbf{J}_c + \mathbf{J}_D = \sigma\mathbf{E} + \frac{\partial}{\partial t}(\varepsilon\mathbf{E}) = \sigma\mathbf{E} + j\omega\varepsilon\mathbf{E} \tag{12.9}$$

From which

$$\frac{\mathbf{J}_c}{\mathbf{J}_D} = \frac{\sigma}{\omega\varepsilon} \tag{12.10}$$

As expected, the displacement current becomes increasingly important as the frequency increases [2].

12.3 FARADAY'S LAW

When a conductor moves through a magnetic field cutting a flux, a voltage is induced in the conductor. Similarly, when the flux cuts across a stationary conductor, a voltage is induced. In either case, the voltage and the rate of cutting the flux are related by Faraday's law:

$$v = -\frac{d\phi}{dt} \tag{12.11}$$

Polarity of the induced voltage is sometimes illustrated by a distorted flux field as shown in Figure 12.3. The lines appear to be pushed ahead of the moving conductor.

A counterclockwise flux is shown around the conductor which would create the same distorted field. By the right-hand rule, the

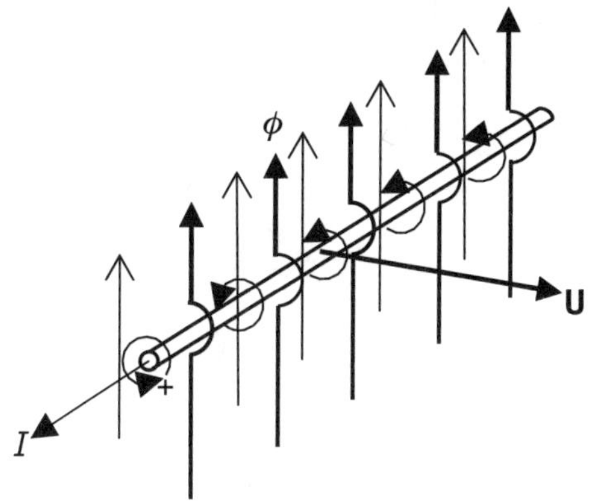

Figure 12.3 Distorted flux field modeling the polarity of the induced voltage.

current, if a closed path were provided, would leave the conductor as shown. Figure 12.4(a) shows the current direction, through an external circuit. Figure 12.4(b) shows an equivalent voltage source and the same current direction through the external circuit.

The negative sign in Faraday's law can be explained by rewriting the equation in integral form.

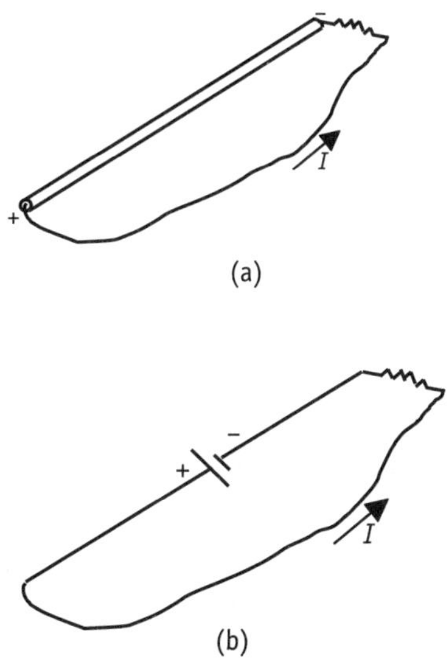

Figure 12.4 (a) Current direction through an external circuit, and (b) equivalent voltage source and same current direction through the external circuit.

$$v = -\frac{d\varphi}{dt} \tag{12.12}$$

$$\oint \mathbf{E} \cdot d\mathbf{l} = -\frac{d}{dt}\int_S \mathbf{B} \cdot d\mathbf{S} \tag{12.13}$$

The contour C around which the integral on the left is taken encloses the plane area S of the surface integral on the right. A positive direction is assigned to C and the direction of the normal, $d\mathbf{S}$, is then determined by the familiar right-hand rule. See Figure 12.5(a).

A changing magnetic flux density, $\mathbf{B}(t)$, is present within the contour. If $\mathbf{B}$ is increasing with time, the time derivative will be positive and, thus, the right side of the equation will be negative. In order for the left integral to be negative, the direction of $\mathbf{E}$ must be opposite to that of the contour as in Figure 12.5(b). A conducting filament in place of the contour would carry a current i_c also in the direction of $\mathbf{E}$.

As shown in Figure 12.5(c), such a current loop generates a flux ϕ' which opposes the increase in $\mathbf{B}$.

Lenz's law summarizes this involved discussion: The voltage induced by a changing flux has a polarity such that the current established in a closed path gives rise to a flux that opposes the change in flux.

In the special case of a conductor moving through a stationary magnetic field, the polarity predicted by Lenz's law will always be such that the conductor experiences magnetic forces that oppose its motion [3].

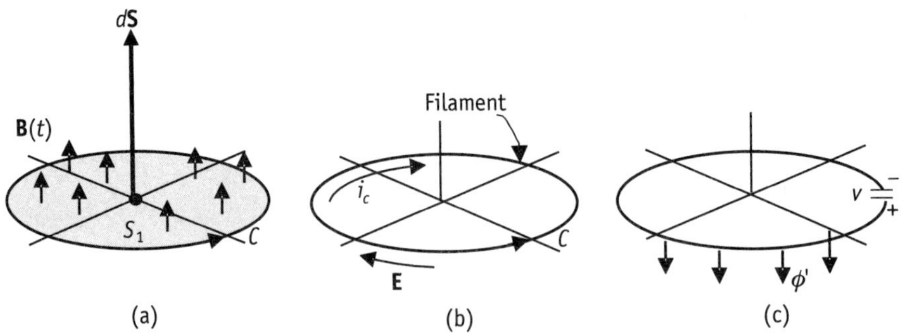

Figure 12.5 (a) Contour C, normal direction $d\mathbf{S}$, (b) direction of $\mathbf{E}$, and (c) current loop generates a flux of ϕ'.

12.4 CONDUCTORS IN MOTION THROUGH TIME-INDEPENDENT FIELDS

The force **F** on a charge Q in a magnetic field **B**, where the charge is moving with velocity **U**, was examined in Chapter 10.

$$\mathbf{F} = Q(\mathbf{U} \times \mathbf{B}) \tag{12.14}$$

A *motional* electric field intensity, $\mathbf{E}_m$, can be defined as the force per unit charge:

$$\mathbf{E}_m = \frac{\mathbf{F}}{Q} = \mathbf{U} \times \mathbf{B} \tag{12.15}$$

When a conductor with a great number of free charges moves through a field, **B**, the impressed $\mathbf{E}_m$ creates a voltage difference between the two ends of the conductor. The magnitude depends on how $\mathbf{E}_m$ is oriented with respect to the conductor. With conductor ends a and b, the voltage of a with respect to b is:

$$v_{ab} = \int_b^a \mathbf{E}_m \cdot d\mathbf{l} = \int_b^a (\mathbf{U} \times \mathbf{B}) \cdot d\mathbf{l} \tag{12.16}$$

If the velocity **U** and the field **B** are at right angles and the conductor is normal to both, then a conductor of length l will have a voltage:

$$v = BlU \tag{12.17}$$

For a closed loop, the line integral must be taken around the entire loop:

$$v = \oint (\mathbf{U} \times \mathbf{B}) \cdot d\mathbf{l} \tag{12.18}$$

Of course, if only part of the complete loop is in motion, it is necessary only that the integral cover this part since $\mathbf{E}_m$ will be zero elsewhere [3].

12.5 CONDUCTORS IN MOTION THROUGH TIME-DEPENDENT FIELDS

When a closed conducting loop is in motion (this includes for loop shape changes) and also the field **B** is a function of time (as well as of position), then the total induced voltage is made up of a contribution from each of the two sources of flux change. Faraday's law becomes:

$$v = -\frac{d}{dt}\int \mathbf{B}\cdot d\mathbf{S} = -\int \frac{\partial \mathbf{B}}{\partial t}\cdot d\mathbf{S} + \oint (\mathbf{U}\times\mathbf{B})\cdot d\mathbf{l} \tag{12.19}$$

The first term on the right is the voltage due to the change in **B**, with the loop held fixed. The second term is the voltage arising from the motion of the loop with **B** held fixed. The polarity of each term is found from the appropriate form of Lenz's law. The two terms are then added with regard to those polarities [4].

12.6 PROBLEMS AND SOLUTIONS[1]

Problem 12.1

Show that the displacement current, i_D in the dielectric of a parallel plate capacitor, shown in Figure 12.6, is equal to the conduction current, i_c, in the leads.

Problem 12.2

In a material for which $\sigma = 5.0$ S/m and $\varepsilon_r = 1$, the electric field intensity is $E = 250\sin(10^{10}t)$ V/m.

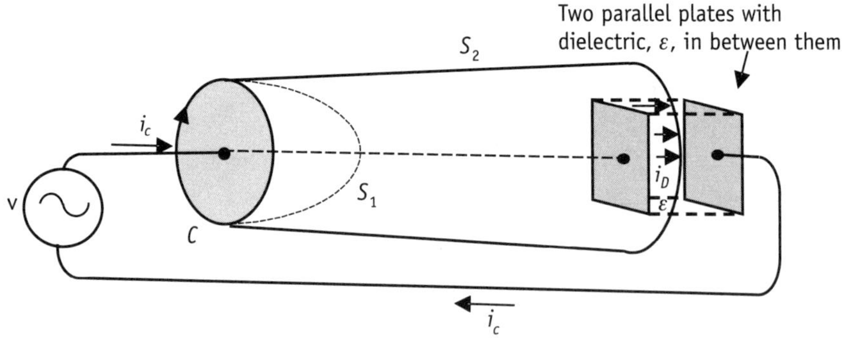

Figure 12.6 Illustration for Problem 12.1.

1. See [2].

Find the conduction and displacement current densities and the frequency at which they have equal magnitudes.

Problem 12.3

A coaxial capacitor with an inner radius of 5 mm, outer radius of 6 mm, and a length of 500 mm has a dielectric $\varepsilon_r = 6.7$ and an applied voltage of 250 sin(377*t*) volts.

Determine the displacement current, i_D. Compare with the conduction current, i_C.

Problem 12.4

In Figure 12.7, a 3m long conductor, parallel to the *x*-axis, moves with a velocity of $\mathbf{U} = 2.5\mathbf{y}$ m/s while in a uniform field of $\mathbf{B} = 0.5\mathbf{z}$ tesla.

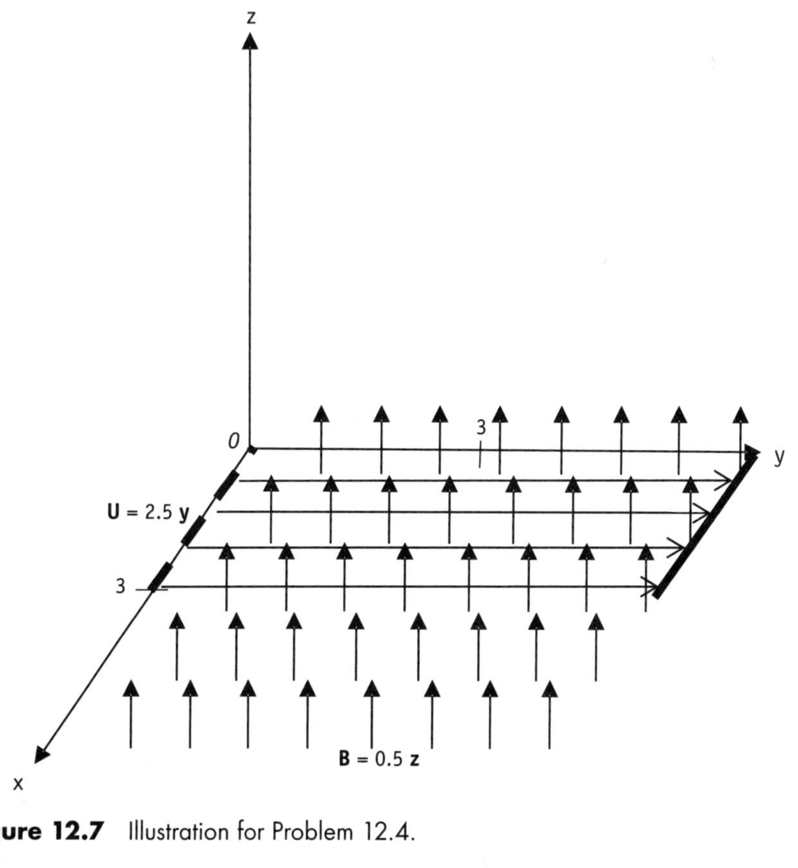

Figure 12.7 Illustration for Problem 12.4.

Find the induced voltage.

Problem 12.5

With the same field and velocity as given in the previous problem, find the voltage induced in the same 3m conductor lying along the y-axis. See Figure 12.8.

Problem 12.6

Find the induced voltage in the conductor in Figure 12.9 where $\mathbf{B} = 0.04\mathbf{y}$ tesla and $\mathbf{U} = 2.5 \sin(10^3\ t)\mathbf{z}$ (m/s).

Problem 12.7

An area of 0.65m^2 in the $z = 0$ plane is enclosed by a filamentary conductor.

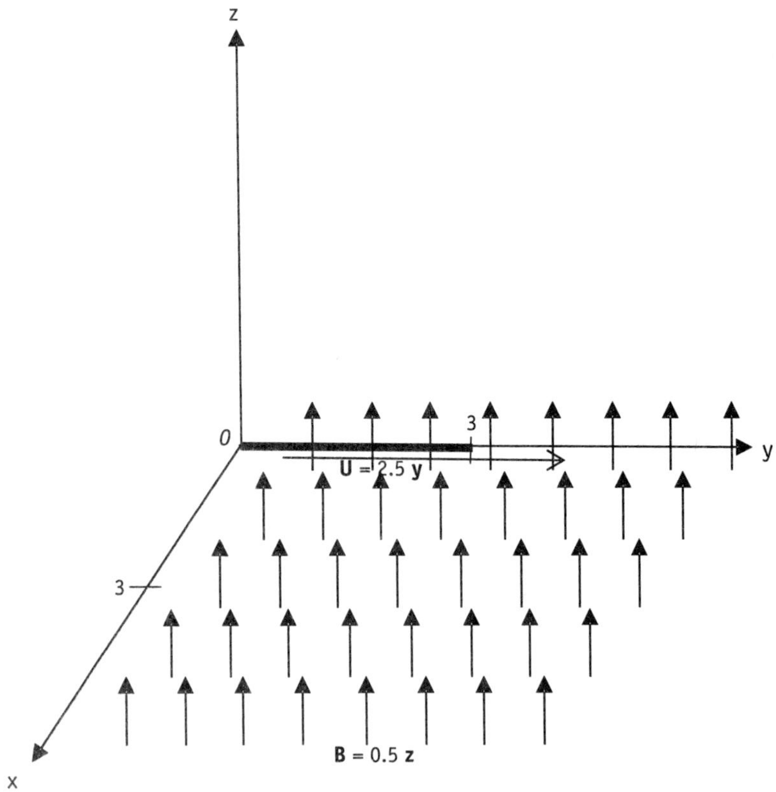

Figure 12.8 Illustration for Problem 12.5.

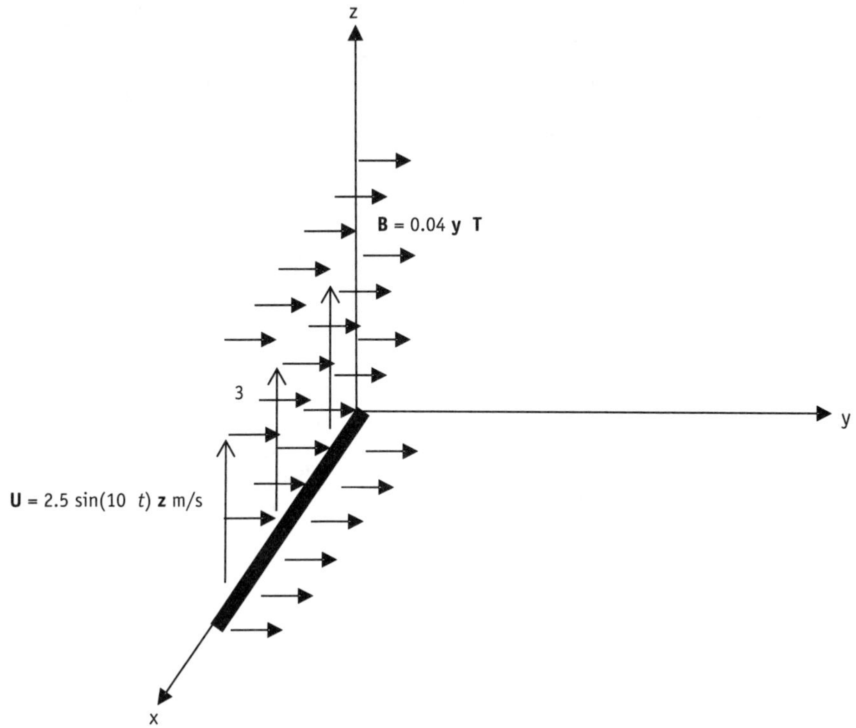

Figure 12.9 Illustration for Problem 12.6.

Find the induced voltage, given the field of:

$$\mathbf{B} = [(0.05/2^{1/2})\cos(10^3 t)]\ (\mathbf{y} + \mathbf{z})\ \text{tesla}$$

See Figure 12.10.

Problem 12.8

The circular loop conductor shown in Figure 12.11 lies in the $z = 0$ plane, has a 0.1m radius, and has a resistance of 5.0Ω.

Given $\mathbf{B} = (0.20)\sin(10^3 t)\mathbf{z}$ tesla, determine the current in the loop.

Problem 12.9

The rectangular loop shown in Figure 12.12 moves toward the origin at a velocity $\mathbf{U} = -250\mathbf{y}$ m/s in a field given by

$$\mathbf{B} = (0.80)\ e^{-0.5y}\mathbf{z}\ \text{tesla}$$

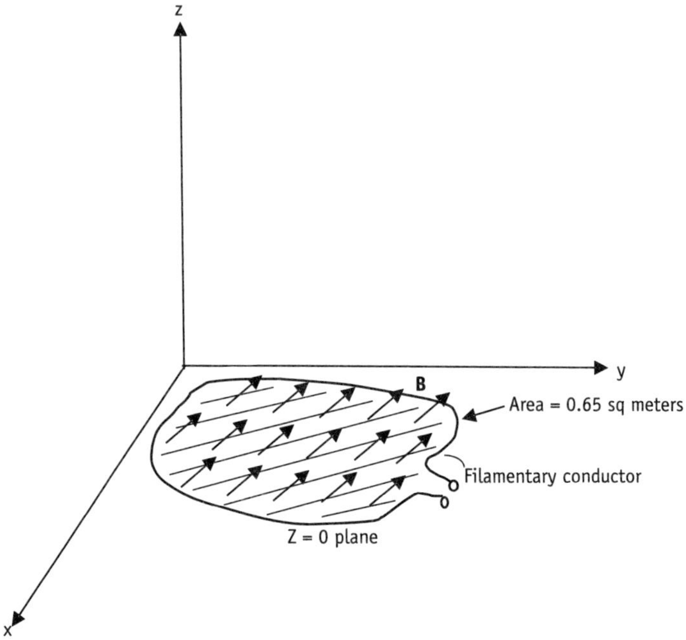

Figure 12.10 Illustration for Problem 12.7.

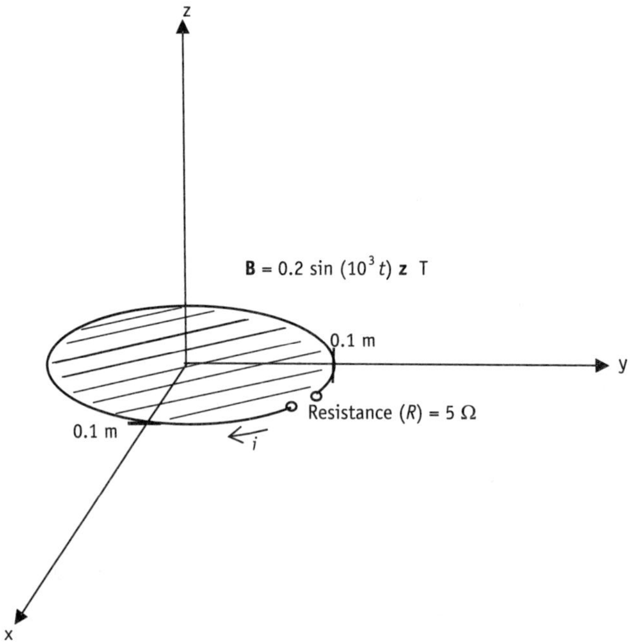

Figure 12.11 Illustration for Problem 12.8.

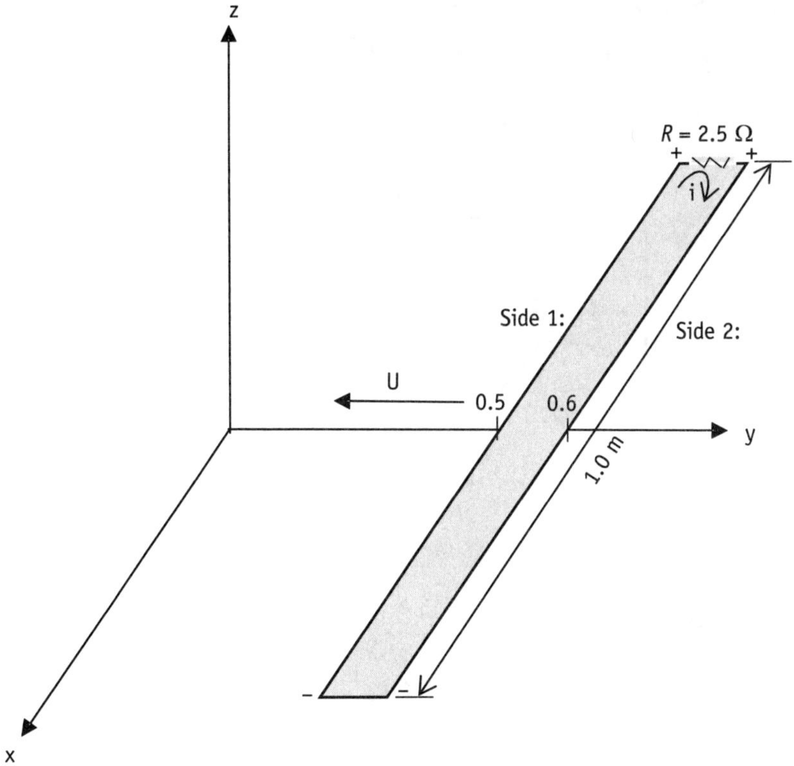

Figure 12.12 Illustration for Problem 12.9.

Find the current at the instant the coil sides are at $y = 0.5$m and 0.6m if $R = 2.5\Omega$.

Problem 12.10

A conducting cylinder of radius 7 cm and height of 15 cm rotates at 600 rpm in a radial field given by $\mathbf{B} = (0.2)\mathbf{r}$ tesla. Sliding contacts at the top and bottom connect to a voltmeter as shown in Figure 12.13.

Find the induced voltage.

Problem 12.11

In Figure 12.14, a rectangular conducting loop with resistance of $R = 0.2\Omega$ turns at a rate of 500 rpm. The vertical conductor a $r_1 = 0.03$m is in a field described as

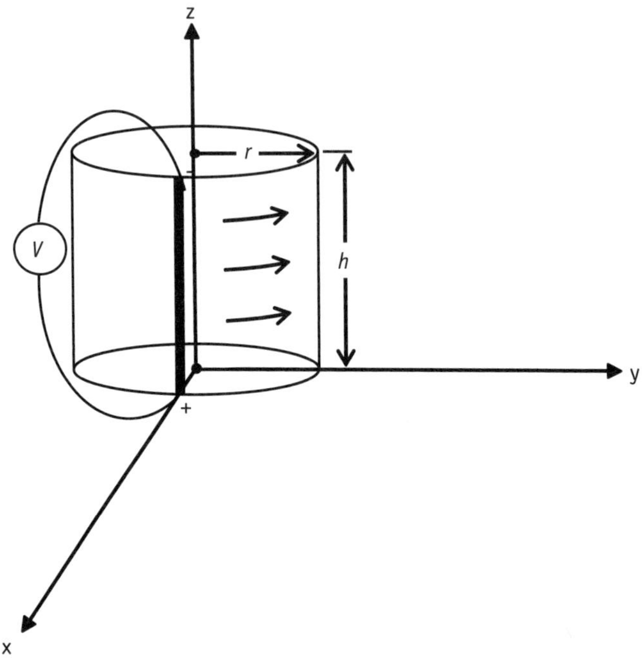

Figure 12.13 Illustration for Problem 12.10.

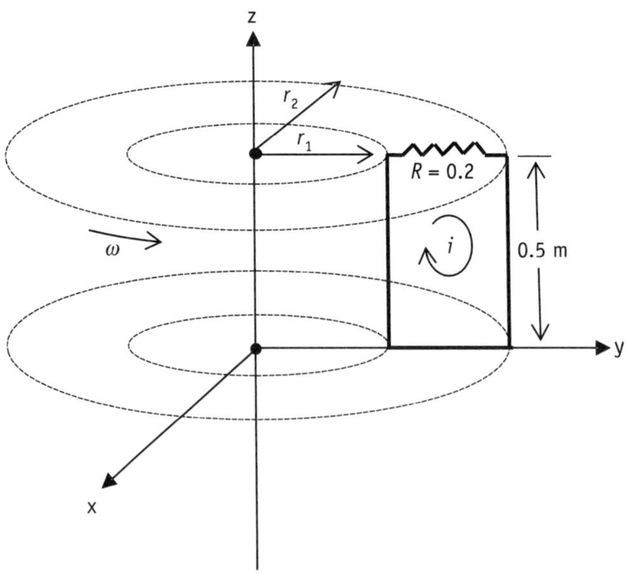

Figure 12.14 Illustration for Problem 12.11.

$$\mathbf{B} = (0.25)\mathbf{r} \text{ tesla}$$

The conductor at $r_2 = 0.05$m is in a field described as

$$\mathbf{B} = (0.8)\mathbf{r} \text{ tesla}$$

Find the current in the loop.

References

[1] Russer, P., *Electromagnetics, Microwave Circuit and Antenna Design for Communications Engineering*, Norwood, MA: Artech House, 2006.

[2] Shadowitz, A., *The Electromagnetic Field*, New York: Dover Publications, 1975.

[3] Warnick, K. F., and P. Russer, *Problem Solving in Electromagnetics, Microwave Circuit and Antenna Design for Communications Engineering*, Norwood, MA: Artech House, 2006.

[4] Cheng, D. K., *Fundamentals of Engineering Electromagnetics*, Reading, MA: Addison-Wesley, 1993.

A

SCIENTIFIC PREFIXES

Prefix	Abbreviation	Decimal Equivalent
Exa	E	10^{18}
Peta	P	10^{15}
Tera	T	10^{12}
Giga	G	10^{9}
Mega	M	10^{6}
Kilo	k	10^{3}
Centi	c	10^{-2}
Milli	M	10^{-3}
Micro	m	10^{-6}
Nano	n	10^{-9}
Pico	p	10^{-12}
Femto	f	10^{-15}
Atto	a	10^{-18}

B

SCIENTIFIC CONSTANTS

e = charge of an electron = 1.602×10^{-19} C

h = Planck's constant = 6.626×10^{-34} Js

k = Boltzmann's constant = 1.380×10^{-23} J/K

T_0 = standard (noise) temperature = 17°C = 290 Kelvin (or K)

$c = v_c$ = velocity of light in a vacuum = 299,792,500 +300 m/s

e = base of natural logarithm system = 2.718

η = free space characteristic impedance – $120\pi\ \Omega = 377\Omega$

ε_0 = dielectric permittivity of free space = 8.854×10^{-12} F/m

μ_0 = magnetic permeability of free space = 1.257×10^{-6} H/m

C

RULES BY WHICH TO PERFORM VECTOR ANALYSIS

C.1 PARALLELOGRAM RULE

Step 1. Parallelogram rule: Connect vectors **A** and **B** in Figure C.1(a) at the tail side of each vector in Figure C.1(b). Connect a third vector **C** tail at the location given in Figure C.1(b). Draw a line parallel to vector **A**, and a second line parallel to vector **B** as shown in Figure C.1(b).

C.2 HEAD-TO-TAIL RULE

Step 1. Head-to-tail rule (**A** + **B**): Examine the lower half of Figure C.1(b) and note vectors **A**, **B**, and **C** connected as depicted in Figure C.2(b).

Step 2. Head-to-tail rule (**B** + **A**): Examine the upper half of Figure C.1(b) and note vectors **A**, **B**, and **C** connected as depicted in Figure C.2(b).

Figure C.1 (a): Two vectors, **A** and **B** (b): parallelogram rule.

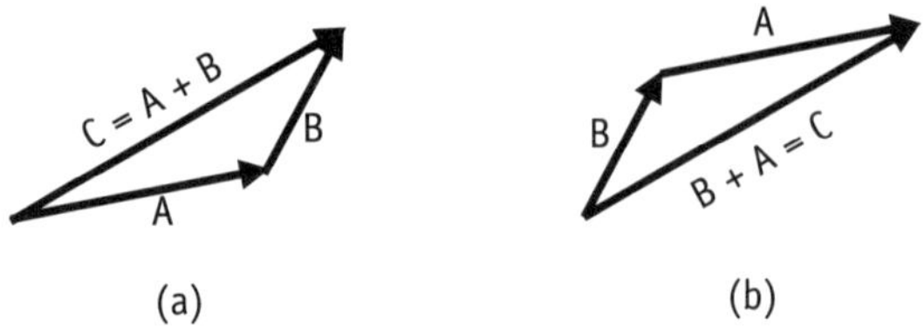

Figure C.2 Head-to-tail rule: (a) Head-to-tail rule, **A** + **B**, (b) head-to-tail rule, **B** + **A**.

Step 3. Perform the respective vector addition as depicted in Figures C.2(a, b).

D

ELECTROMAGNETIC SPECTRUM AND FREQUENCY BAND DESIGNATIONS

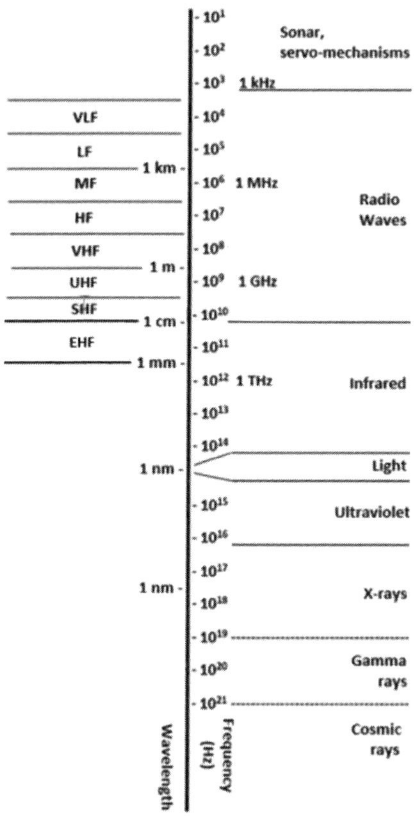

Figure D.1 Electromagnetic spectrum.

Table D.1
Frequency Band Designations

Band Name	Frequency Range
UHF	0.3–1.0 GHz
L-band	1.0–1.5 GHz
S-band	1.5–3.9 GHz
C-band	3.9–8.0 GHz
X-band	8.0–12.0 GHz
Ku-band	12.5–18.0 GHz
K-band	18.0–26.5 GHz
Ka-band	26.5–40.0 GHz
U-band	40.0–60.0 GHz
V-band	50.0–75.0 GHz
W-band	75.0–100.0 GHz

E

TRANSMISSION LINE EQUATIONS, GENERAL LINE EXPRESSIONS, AND IDEAL LINE EXPRESSIONS

Table E.1
Transmission Line Equations, General Transmission Line Expressions, and Ideal Line Expressions

Quantity	**General Line Expression**	**Ideal Line Expression**
Propagation constant	$\gamma = \alpha + j\beta$	$\gamma = j\omega(LC)^{1/2}$
Phase constant, β	Imaginary Part of γ	$\beta = \omega(LC)^{1/2} = 2\pi/\lambda$
Attenuation constant, α	Real Part of γ	0
Characteristic impedance, Z_0	$Z_0 = \left[\frac{(R + j\omega L)}{(G + j\omega C)}\right]^{1/2}$	$Z_0 = \left(\frac{L}{C}\right)^{1/2}$
Input impedance	$Z_{in} = Z_0\left[\frac{(Z_L + Z_0 \tanh(\gamma l))}{(Z_0 + Z_L \tanh(\gamma l))}\right]^{1/2}$	$Z_{in} = Z_0\left[\frac{(Z_L + jZ_0 \tan(\beta l))}{(Z_0 + jZ_L \tan(\beta l))}\right]^{1/2}$
Impedance of short-circuited line, $Z_L = 0$	$Z_{s.c.} = Z_0 \tanh(\gamma l)$	$Z_{s.c.} = jZ_0 \tan(\beta l)$
Impedance of open-circuited line, $Z_L = \infty$	$Z_{o.c.} = Z_0 \coth(\gamma l)$	$Z_{o.c.} = -jZ_0 \cot(\beta l)$
Impedance of line, odd no. of ½ λs long	$Z = Z_0\left[\frac{(Z_L + Z_0 \coth(\alpha l))}{(Z_0 + Z_L \coth(\alpha l))}\right]$	$Z = \frac{Z_0^2}{Z_L}$
Impedance of line, integer no. of ½ λs long	$Z = Z_0\left[\frac{(Z_L + Z_0 \tanh(\alpha l))}{(Z_0 + Z_L \tanh(\alpha l))}\right]$	$Z = Z_L$
Voltage reflection coefficient	$\Gamma = \frac{(Z_L - Z_0)}{(Z_L + Z_0)}$	$\Gamma = \frac{(Z_L - Z_0)}{(Z_L + Z_0)}$

Table E.2
Helpful Transmission Line-Related Equations

Equation	Explanation
$r = \frac{1+\|\Gamma\|}{1+\|\Gamma\|}$ $\Gamma = \frac{(r-1)}{(r+1)}$	R = VSWR $\|\Gamma\|$ = magnitude of reflection coefficient
$\Gamma = \frac{(R - Z_0)}{(R + Z_0)}$ $r = \frac{R}{Z_0}$ $r = \frac{Z_0}{R}$	Γ = reflection coefficient (real) at a point in a line where impedance is real (R) $R > Z_0$ (at voltage maximum) $Z_0 < R$ (at voltage minimum)
$\frac{P_t}{P_i} = \|\Gamma\|^2 = \left(\frac{r-1}{r+1}\right)^2$ $\frac{P_t}{P_i} = 1 - \|\Gamma\|^2 = \frac{4r}{(r+1)}$	P_t = transmitted power P_r = reflected power P_i = incident power

F

MAXWELL'S EQUATIONS

F.1 FREE SPACE SET

Table F.1

Maxwell's Equations, Free Space Set: Point Form Versus Integral Form

Point Form	**Integral Form**
$\nabla \times \bar{\mathbf{H}} = \dfrac{\partial \bar{\mathbf{A}}}{dt}$	$\oint \bar{\mathbf{H}} \ast d\mathbf{l} = \int_S \left(\dfrac{\partial \bar{\mathbf{D}}}{\partial t} \right) \ast d\bar{\mathbf{S}}$
$\nabla \times \bar{\mathbf{E}} = -\dfrac{\partial \bar{\mathbf{B}}}{dt}$	$\oint \bar{\mathbf{H}} \ast d\mathbf{l} = \int_S \left(\dfrac{\partial \bar{\mathbf{D}}}{\partial t} \right) \ast d\bar{\mathbf{S}}$
$\nabla \ast \bar{\mathbf{D}} = 0$	$\oint_S \bar{\mathbf{D}} \ast d\bar{\mathbf{S}} = 0$
$\nabla \times \bar{\mathbf{B}} = 0$	$\oint_S \bar{\mathbf{B}} \ast d\bar{\mathbf{S}} = 0$

F.2 GENERAL SET

Table F.2
Maxwell's Equations, General Set: Point Form Versus Integral Form

Point Form	Integral Form	
$\nabla \times \bar{\mathbf{H}} = \bar{\mathbf{J}}_c + \frac{\partial \bar{\mathbf{D}}}{dt}$	$\oint \bar{\mathbf{H}} * d\mathbf{l} = \int_S \left(\frac{\partial \bar{\mathbf{D}}}{\partial t} \right) * d\bar{\mathbf{S}}$	(Ampere's law)
$\nabla \times \bar{\mathbf{E}} = -\frac{\partial \bar{\mathbf{B}}}{dt}$	$\oint \bar{\mathbf{H}} * d\mathbf{l} = \int_S \left(\frac{\partial \bar{\mathbf{D}}}{\partial t} \right) * d\bar{\mathbf{S}}$	(Faraday's law; *S* fixed)
$\nabla * \bar{\mathbf{D}} = \rho$	$\oint_S \bar{\mathbf{D}} * d\bar{\mathbf{S}} = \int_v \rho \, dv$	(Gauss' law)
$\nabla * \bar{\mathbf{B}} = 0$	$\oint_S \bar{\mathbf{B}} * d\bar{\mathbf{S}} = 0$	(Nonexistence of magnetic monopole)

ACRONYMS AND ABBREVIATIONS

ADF	Automatic direction finding
BCA	Boeing Commercial Airplanes
EE	Electrical engineer/engineering
EM	Electromagnetic
EME	Electromagnetic effects
FAA	Federal Aviation Administration
FSL	Free-space loss
FZL	Fresnel zone loss
GHz	Gigahertz
HF	High frequency
IEEE	Institute of Electrical/Electronic Engineers
IL	Insertion loss
LOS	Line-of-sight
MHz	Megahertz
MPL	Multipath loss
MW	Microwave frequency
NDB	Nondirectional beacon
RADAR	Radio detection and ranging
RF	Radio frequency

SINAD	Signal plus noise to noise and distortion
SNR	Signal-to-noise ratio
VHF	Very high frequency
VOR	VHF omnidirectional ranging

SELECTED BIBLIOGRAPHY

Blake, L. V., *Antennas*, Dedham, MA: Artech House, 1984.

Brown, R. G., et al., *Lines, Waves, and Antennas: The Transmission of Electric Energy*, Second Edition, New York: John Wiley and Sons, 1973, Chapters 12–15.

Budlong, A. L., *The A. R. R. L. Antenna Book*, 9th Edition, West Hartford, CT: The American Radio Relay League, 1960.

De Martino, A., *Introduction to Modern EW Systems*, Norwood, MA: Artech House, 2012, Appendix C.

Elbert, B. R., *The Satellite Communication Applications Handbook*, Norwood, MA: Artech House, 1997.

Hecht, E., *Schaum's Outline of Theory and Problems of Optics*, New York: McGraw-Hill, 1975.

Howard W. Sams and Co., *Reference Data for Radio Engineers*, 6th Edition, Indianapolis, IN: ITT, 1976.

Howard W. Sams and Co., *Reference Data for Radio Engineers*, 5th Edition, Indianapolis, IN: ITT, 1972.

Johnson, R. C. (ed.), *Antenna Engineering Handbook*, 3rd Edition, New York: McGraw-Hill, 1993.

Kennedy, G., *Electronic Communications Systems*, 2nd Edition, New York: McGraw-Hill, 1977.

Macnamara, T., *Handbook of Antennas for EMC*, Norwood, MA: Artech House, 1995.

Poisel, R. A., *Antenna Systems and Electronic Warfare Applications*, Norwood, MA: Artech House, 2012.

Russer, P., *Electromagnetics, Microwave Circuit and Antenna Design for Communications Engineering*, 2nd Edition, Norwood, MA: Artech House, 2006.

Saakian, A., *Radio Wave Propagation Fundamentals*, Norwood, MA: Artech House, 2011.

Shadowitz, A., *The Electromagnetic Field*, Mineola, NY: Dover Publications, 1975.

Stutzman, W. L., and G. A. Thiele, *Antenna Theory and Design*, New York: John Wiley and Sons, 1981.

Volakis, J. L. (ed.), *Antenna Engineering Handbook*, 4th Edition, McGraw-Hill, 2007.

Warnick, K. F., and P. Russer, *Problem Solving in Electromagnetics, Microwave Circuit and Antenna Design for Communications Engineering*, Norwood, MA: Artech House, 2006.

Wiley, R. G., *Electronic Intelligence: The Analysis of Radar Signals*, 2nd Edition, Norwood, MA: Artech House, 1993, Chapters 4 and 5.

ABOUT THE AUTHOR

Dean James Friesen graduated from Kansas State University (Manhattan, KS), College of Engineering in 1987 with a BS degree in engineering technology (electrical/electronic emphasis) and in 1984 with an associates of technology (AT) degree in electronic technology from Kansas State Technical Institute (now Kansas State Polytechnic, Salina, KS). Mr. Friesen worked for 36 years in the EM-related engineering-related disciplines of EW, RF/MW/mm-wave, HEMP, EMC/EMI, E3, and EME. He retired from The Boeing Company in June 2020 as the Boeing Commercial Airplanes (BCA), Product Development, EME and Antennas Group, senior principal investigator (2016–2020).

Mr. Friesen began his military defense engineering career at Boeing-Wichita in Kansas (B52 defensive systems) in 1987 where he performed B52H electronic warfare (EW) work for the U.S. Air Force for 2-1/2 years. In 1990, he joined the International Space Station (ISS) project (formerly called Space Station Freedom) at Boeing-Huntsville in Alabama for almost 2 years. Mr. Friesen's desire to return to U.S. defense engineering moved him to join CAS, Inc. (Patriot Weapon System analysis, Huntsville, AL) in 1991. He also worked for TX Instruments (defense sector, RF/MW Group, Dallas, TX) in 1996, Northrop-Grumman Corporation (Land Combat Systems (LCS), Huntsville, AL) in 1999, and Boeing-Wichita (Boeing Defense Systems) again in 2003.

The untimely closing of the Boeing-Wichita plant in 2012 caused Mr. Friesen to join BCA to work in the Puget Sound area (Everett/ Renton/Mukilteo in Washington state) for 8 years. He worked the first four years as the 737MAX airplane BCA EME team engineering lead. In 2016, he was recruited specifically to join the BCA Product Development organization, EME and Antennas Group, as a principal investigator where he was promoted in 3 months to be the group's sole senior principal investigator for 4 years before retiring in June 2020.

Post-retirement, Mr. Friesen was called back into service in 2021 to work as a senior technical writer for Boeing Design Practices for a 16-month engineering labor service contract period. He was again called back into BCA EME engineering in 2022 to assist EME engineering-unit managers with writing and researching information. In 2023, he was also asked to help with the EME side of requirement management (validation). As of 2023, Mr. Friesen has been asked to help lead and train new-hire EME engineers in the subject of electromagnetic effects as applied to commercial airplanes.

Mr. Friesen has been married for 31 years and has three children and nine grandchildren.

INDEX

E

F

G

H

I

W

Z

Artech House
Electromagnetic Analysis Library

Christos Christodoulou, Series Editor
Vince Rodriguez, Series Editor

Advanced Computational Electromagnetic Methods and Applications, Wenhua Yu, Wenxing Li, Atef Elsherbeni, Yahya Rahmat-Samii, editors

Advanced FDTD Methods: Parallelization, Acceleration, and Engineering Applications, Wenhua Yu, et al.

Advances in FDTD Computational Electrodynamics: Photonics and Nanotechnology, Allen Taflove, editor

Analysis Methods for Electromagnetic Wave Problems, Volume 1, Eikichi Yamashita, editor

Analysis Methods for Electromagnetic Wave Problems, Volume 2, Eikichi Yamashita, editor

Analytical and Computational Methods in Electromagnetics, Ramesh Garg

Analytical Modeling in Applied Electromagnetics, Sergei Tretyakov

Anechoic Range Design for Electromagnetic Measurements, Vince Rodriguez

Applications of Neural Networks in Electromagnetics, Christos Christodoulou and Michael Georgiopoulos

CFDTD: Conformal Finite-Difference Time-Domain Maxwell's Equations Solver, Software and User's Guide, Wenhua Yu and Raj Mittra

Computational Electrodynamics: The Finite-Difference Time-Domain Method, Second Edition, Allen Taflove and Susan C. Hagness

Computational Electromagnetics, Prabha Umashankar and Allen Taflove

Electromagnetic Fields in Multilayered Structures: Theory and Applications, Arun Bhattacharyya

Electromagnetics for Engineers, Volume 1: Electrostatics and Magnetostatics, Dean James Friesen

Electromagnetic Modeling of Composite Metallic and Dielectric Structures, Branko M. Kolundzija and Antonije R. Djordjevic

Electromagnetic Scattering and Material Characterization, Abbas Omar

Electromagnetic Waves in Chiral and Bi-Isotropic Media, I. V. Lindell, et al.

Electromagnetic Diffraction Modeling and Simulation with MATLAB®, Gökhan Apaydin and Levent Sevgi

Engineering Applications of the Modulated Scatterer Technique, Jean-Charles Bolomey and Fred E. Gardiol

Far-Field Wireless Power Transfer and Energy Harvesting, Naoki Shinohara and Jiafeng Zhou, editors

Fast and Efficient Algorithms in Computational Electromagnetics, Weng Cho Chew, et al., editors

FDTD Modelling of Metamaterials: Theory and Applications, Yang Hao and Raj Mittra

Generalized Multipole Technique, Christian Hafner

Grid Computing for Electromagnetics, Luciano Tarricone and Alessandra Esposito

High Frequency Electromagnetic Dosimetry, David A. Sánchez-Hernández, editor

High-Power Electromagnetic Effects on Electronic Systems, D. V. Giri, Richard Hoad, and Frank Sabath

Integral Equation Methods for Electromagnetics, Joseph R. Mautz and Nagayoshi Morita

Intersystem EMC Analysis, Interference, and Solutions, Uri Vered

Introduction to Electromagnetic Wave Propagation, Paul Rohan

Introduction to the Uniform Geometrical Theory of Diffraction, Derek A. McNamara, C. W. I. Pistorius, and Jan Malherbe

Iterative and Self-Adaptive Finite-Elements in Electromagnetic Modeling, Magdalena Salazar-Palma, et al.

Machine Learning Applications in Electromagnetics and Antenna Array Processing, Manel Martínez-Ramón, Arjun Gupta, José Luis Rojo-Álvarez, and Christos Christodoulou

New Foundations for Applied Electromagnetics: The Spatial Structure of Fields, Said Mikki and Yahia Antar

Numerical Analysis for Electromagnetic Integral Equations, Karl F. Warnick

Parallel Finite-Difference Time-Domain Method, Wenhua Yu, et al.

Practical Applications of Asymptotic Techniques in Electromagnetics, Francisco Saez de Adana, et al.

A Practical Guide to EMC Engineering, Levent Sevgi

Quick Finite Elements for Electromagnetic Waves, Second Edition, Giuseppe Pelosi, Roberto Coccioli, and Stefano Selleri

The Scattering of Electromagnetic Waves from Rough Surfaces, Peter Beckman and Andre Spizzichino

Shipboard Electromagnetics, Preston Law

Understanding Electromagnetic Scattering Using the Moment Method: A Practical Approach, Randy Bancroft

Wavelet Applications in Engineering Electromagnetics, Tapan K. Sarkar, Magdalena Salazar-Palma, and Michael C. Wicks

For further information on these and other Artech House titles, including previously considered out-of-print books now available through our In-Print-Forever® (IPF®) program, contact:

Artech House Publishers	Artech House Books
685 Canton Street	16 Sussex Street
Norwood, MA 02062	London SW1V 4RW UK
Phone: 781-769-9750	Phone: +44 (0)20 7596 8750
Fax: 781-769-6334	Fax: +44 (0)20 7630 0166
e-mail: artech@artechhouse.com	e-mail: artech-uk@artechhouse.com

Find us on the World Wide Web at: www.artechhouse.com